"Jess Fanzo has seen [...] and everything, and talked to every one. You can have no better, more knowledgeable guide to the mess the food system is in, and how we can get out of it. If you want to do something about the global nutrition crisis, read her book, and roll up your sleeves, as she does."

— Luigi Guarino, Director of Science,
Global Crop Diversity Trust

"Jessica Fanzo argues that dinner not only can fix the planet, but must. Read her book. It's beautifully written, authoritative, and utterly convincing—essential reading for anyone interested in the world's food problems."

— Marion Nestle, professor emerita, New York University,
and author of *Let's Ask Marion: What You Need to Know
about the Politics of Food, Nutrition, and Health*

"Dr. Fanzo's book is not only a wake-up call for eaters, but a road-map for how to make our global food system more ecologically and socially just. She explains the fragility of our current way of producing food, while giving us hope that things can change for the better."

— Danielle Nierenberg, President of Food Tank
and 2020 Julia Child Award Recipient

"Your food choices truly matter for both your own and our environment's health. *Can Fixing Dinner Fix the Planet?* guides you through the far-reaching impacts of your decisions, and provides hands-on ways to combat the issues we face. For the love of food and the planet, dig in, and bon appetit!"

— Michiel Bakker, Vice President,
Google Global Workplace Services Programs

"We've never needed to be more aware of the impacts of our food choices—or to listen more carefully to pioneering experts like Jess Fanzo. Written with wit, insight and a real sense of urgency, this is essential reading for anyone with a personal or professional stake in what they eat and where it's sourced from."

— Gunhild Stordalen, Founder and Executive Chair, EAT Foundation

"A unique synthesis that weaves together revealing data with the author's personal experience, *Can Fixing Dinner Fix the Planet?* demonstrates the negative impacts food systems are having on health and the environment. Fanzo's description of her work in some of the hardest-hit communities reveals how the global challenges of providing healthy and sustainable diets for all leaves no region untouched. Readers of Michael Pollan, Mark Bittman, Frances Moore Lappé, and Marion Nestle will be interested in this nuanced book."

— Michael Clark, Nuffield Department of Population Health,
University of Oxford

Can Fixing Dinner Fix the Planet?

JOHNS HOPKINS
WAVELENGTHS

In classrooms, field stations, and laboratories in Baltimore and around the world, the Bloomberg Distinguished Professors of Johns Hopkins University are opening the boundaries of our understanding on many of the world's most complex challenges. The Johns Hopkins Wavelengths series brings readers inside their stories, presenting the pioneering discoveries and innovations that benefit people in their neighborhoods and across the globe in artificial intelligence, cancer research, food systems, health equity, science diplomacy, and other critical areas of study. Through these compelling narratives, their insights will spark conversations from dorm rooms to dining rooms to boardrooms.

This print and digital media program is a partnership between the Johns Hopkins University Press and the University's Office of Research. Team members include:

Consultant Editor: Sarah Olson

Senior Acquisition Editor: Matthew R. McAdam

Copyeditor: Andre M. Barnett

Art Director: Martha Sewall

Designer: Matthew Cole

Production Manager: Jennifer Paulson

Publicist: Rebecca Rozenberg

Program Manager: Anna Marlis Burgard

JHUP Director and Publisher: Barbara Kline Pope

Office of Research Executive Director for Research: Julie Messersmith

Can Fixing Dinner Fix the Planet?

JESSICA FANZO, PhD

Johns Hopkins University Press
Baltimore

Johns Hopkins Wavelengths is a trademark of the Johns Hopkins University.

© 2021 Johns Hopkins University Press
All rights reserved. Published 2021
Printed in the United States of America on acid-free paper
9 8 7 6 5 4 3 2 1

Johns Hopkins University Press
2715 North Charles Street
Baltimore, Maryland 21218-4363
www.press.jhu.edu

Library of Congress Cataloging-in-Publication Data

Names: Fanzo, Jessica, author.
Title: Can fixing dinner fix the planet? / Jessica Fanzo.
Description: Baltimore : Johns Hopkins University Press, 2021. |
   Series: Johns Hopkins wavelengths | Includes bibliographical
   references and index.
Identifiers: LCCN 2020041842 | ISBN 9781421441122 (hardcover) |
   ISBN 9781421441139 (ebook) | ISBN 9781421441146 (ebook open access)
Subjects: LCSH: Food habits—Health aspects—Longitudinal studies. |
   Food habits—Environmental aspects—Longitudinal studies. | Food
   supply—Environmental aspects—Longitudinal studies.
Classification: LCC GT2850 .F36 2021 | DDC 394.1/2—dc23
LC record available at https://lccn.loc.gov/2020041842
A catalog record for this book is available from the British Library.

*Special discounts are available for bulk purchases of this book. For more information,
please contact Special Sales at specialsales@jh.edu.*

Johns Hopkins University Press uses environmentally friendly book materials,
including recycled text paper that is composed of at least 30 percent post-consumer
waste, whenever possible.

Contents

Preface

AT THE BEGINNING OF MY CAREER, I never would have guessed that I'd end up where I am today. I was a lab rat. My bachelor's, master's, and PhD degrees were all focused on nutrition, but at the molecular level. For many years I sat at benches pipetting liquids into tubes, examining the interactions between genes and nutrients. Siloed in ivory towers, I contemplated the esoteric fine points of nutritional biochemistry, oblivious to what was happening on the streets of the world.

After my PhD work in molecular nutrition and following my postdoctoral fellowship in immunology, I wanted to focus more on people to see more immediate outcomes. So, I left "bench" science and worked for a time at the Doris Duke Charitable Foundation on issues of global public health. Elaine Gallin, the director of Medical Research at that time, took me under her wing and exposed me to the many experts doing cutting-edge research in global health, particularly HIV/AIDS, tuberculosis, and malaria. We traveled to Africa—my first time on the continent—and got to see firsthand how HIV was ravaging South Africa and Uganda. I hung up my lab coat for good.

After my years at Doris Duke, I began working with world-renowned Professor Jeffrey Sachs, an expert on economic development and poverty, and his large team at Columbia University's Earth Institute. Eventually, I relocated to Kenya with my partner, where I served as the nutrition regional adviser for East and Southern Africa at the Millennium Development Goal Centre, working on both nutrition policy design with governments and implementing programs in local villages. I learned to think more broadly about how nutrition fits into sustainable development and where and how it links to other areas such as agriculture, economics, water, environment, gender dynamics, and health. I had the opportunity to work with committed and knowledgeable experts in those fields, and together we put these disciplines to work for international development. I was introduced to people such as World Food Prize winner Pedro Sanchez, who opened up the world of agriculture to me. Cheryl Palm and Glenn Denning taught me how critical the environment and ecosystems are for food security and thus national security. Working at the interface of research, policy, and practice, I began to see the fuller spectrum of influences and effects nutrition can make on people's health, livelihoods, and well-being.

After working deeply in Africa for about a decade, I began research in Asia as well, in Timor-Leste, Nepal, and Myanmar. This is when I really discovered my love for "boots on the ground" fieldwork, far removed from the sterile labs of my earlier years. Much of what I've learned comes from conversations with farmers, mothers, fathers, and their children, as well as the students and postdoctoral fellows, who work tirelessly on our global research portfolio. I'm humbled and grateful to have spent time with all of them.

After homing in on specific areas of Africa and Asia, I transitioned into working more globally on food system challenges, having taken up the position as an assistant professor at Columbia University as well as holding posts with Bioversity International, the REACH Partnership at the UN World Food Programme, and the Food and Agriculture Organization of the United Nations. Working with so many stakeholders in our food systems gave me new perspective and insights into how nutrition links to climate change, economic growth, and overall sustainable development. I had the opportunity to become a part of important global commissions and publications that have informed the field—and this book—including the Global Nutrition Report, the High Level Panel of Experts on Food Systems, and the EAT-*Lancet* Commission, all of which will be mentioned later. In 2012, I was honored with the Premio Daniel Carasso Prize, validating and encouraging my work on sustainable food and diets for long-term human health.

In 2015, I became the eleventh Bloomberg Distinguished Professor at Johns Hopkins University. I collaborate within three different schools—the Nitze School of Advanced International Studies (SAIS), the Berman Institute of Bioethics, and the Department of International Health of the Bloomberg School of Public Health. Johns Hopkins is an amazingly integrative and cutting-edge place. I'm privileged to connect and team up with some of the world's preeminent epidemiologists, ethicists, and political scientists on grand challenges across food systems.

For much of my career I've investigated the complex interactions among food systems (everything involving food, from farm to fork), diets, human health, and the climate crisis. In particular, I've studied how food systems could be changed to promote healthy, sustainable, and equitable diets. Along the way, I've learned a lot about the actions we must take as individuals and as members of local, national, and international communities to ensure the future health of humanity and our planet. This book lays out what I've learned from experiences around the world, analyzes the problems we face, and offers solutions that I'm convinced can solve those problems. Every society cares about food—it's the lifeblood that shapes individual health and vast cultures on a daily basis. But without the right amount or quality of foods to eat, things can go very wrong, very quickly, especially when shocks to the systems occur, brought on by armed conflicts, droughts, and other extreme environmental and human events. And the decisions about these foods—from how they're raised, how far they travel to get to stores, and how much packaging surrounds them—impact our planet in profound ways, from its physical environment to what it supports: the millions of plants, insects, and animals, including us.

While this book was being written, the SARS-CoV-2 virus that causes COVID-19 spread across the globe like wildfire, giving few clues to where we stand within the pandemic. Are we at the beginning, in the middle, or at the end of its wrath? What's become apparent is that what began as a food system–related zoonotic (spread between animals and humans) disease shook the global health system to its core and has ramifications in every other sector, too, including the worldwide food and financial systems. Human activity is the biggest instigator of change in animal-human interactions, and much of that has to do with agriculture. No other species has so profoundly changed the planet and the ecosystems that support species' diversity in such a short span of time.

As COVID-19 spread from person to person, community to community, and nation to nation, it illustrated just how interconnected we all are—how what happens to one person can impact thousands, even millions, while also shedding light on how ultimately fragile a massive engine like the international food supply can be. If the near-term food insecurity and hunger fallouts aren't addressed, any actions could stymie progress in tackling COVID-19, not only in the present time but as the pandemic continues to spread and mutate around the world over the next one to three years. Estimates by the UN World Food Programme project that the number of people facing acute food insecurity will rise significantly as a result of the economic impact of COVID-19.

Obesity and other noncommunicable diseases are considered significant risk factors for COVID-19 hospitalizations and we're finding there could be serious complications even for those who are asymptomatic. COVID-19 directly challenges the ability to access healthy foods because of the shortcomings of the global food system, such as inefficient and inequitable distribution of food and the inadequate attention to food system workers such as those who work in meat processing centers.

For some, cooking and eating are about basic survival, while for others, it's a pleasurable pastime, even an art form. What I've learned over the years, and what the COVID-19 pandemic has shown me, is that we're inextricably bound together by food. Scores of cooking and dining shows on television present meal preparation not only as a means to a great dining experience but also as entertainment, with colorful hosts and competitions. We spend a good portion of our days considering, shopping for, cooking, and eating food and cleaning up what remains. In some parts of the world, eating still involves walking some distance to get water and growing or raising what's eaten. All of our collective actions and decisions have ripple effects across countries and, often, around the world.

In the essay "Goodbye to All That," Joan Didion wrote: "It is easy to see the beginnings of things and harder to see the ends." When we think about the food system and where we sit right now in the world, it is indeed harder to see where it all will end. There's no program yet for what the local and global impacts of our food choices are or how the planet will respond and in turn shape us. When COVID-19 is in the rearview mirror, what will we look like as a human society? Will we be better informed, ready, and more resilient before the next pandemic or climate shock strikes? I'm hopeful, and I trust that human perseverance, creativity, and ingenuity will pull us through to the other side.

The food security challenges we face are not trivial. As global citizens, we are at a critical juncture amid the perils of climate change, pandemics, and political upheaval. Within the swirling chaos, the opportunities for equitable, healthy, and sustainable food systems are substantial but will require that high-quality science be translated into policy faster than ever before. I'm optimistic in light of the many scientists and inventors around the world who are helping course correct the problems we face, putting us on the right track. Research can bring wholesale changes to action and politics. Right this minute, many researchers are working tirelessly in field stations, farms, labs, conference rooms, and classrooms to establish clear understandings of factors that feed the problems of global food systems and institute solutions to be taken up by individuals, organizations, private companies, and nations. This book is, in a way, a thank you to them and to the many scientists who contributed their thoughts and their research to the concepts laid out in the following pages. I cannot thank them enough for shaping my thinking on these challenging issues; I hope their work will raise your awareness, and inspire you to carefully consider your own choices.

We have gaps in our understanding about how to ensure that food systems might be sustainable, equitable, and healthy for everyone, leaving no one behind. Researchers and scientists must have a voice and dutifully fill in those gaps the world makes. Politicians, business owners, and citizens of the world must then do their part to help. This is our chance to design and construct the observable ending and move forward toward a more sustainable world, to coexist in accord with the planet while we nourish its citizens.

Can Fixing Dinner
Fix the Planet?

Yes, We'll Have No Bananas

THE NEXT TIME YOU'RE IN A GROCERY STORE, take a minute to think about the bananas in the produce section. While more than a thousand varieties are consumed locally around the world—grown in tropical nations such as the world's top exporter, Ecuador—most of the bananas shipped to Europe and the United States are a single genetically modified variety known as Cavendish bananas, which are seedless, sweeter, and have higher yields and thicker peels that resist damage during harvesting and shipping.

On plantations, bananas are picked, washed, and packaged. They then travel in refrigerated cargo ships to distant ports, burning untold gallons of fossil fuels along the way. At their destinations, they're treated with gases in temperature-controlled warehouses to trigger ripening. After a health and safety inspection, they're trucked to retail outlets for sale, further increasing their carbon footprints.

The Cavendish's ancestry as a genetically modified organism has made it vulnerable to a soil-borne fungus; in the 1950s, the previous commercial banana superstar, the Gros Michel, was similarly wiped out by a fungus in what was akin to a plant-world pandemic. Much like human ailments that can no longer be effectively treated by available antibiotics, such fungi are starting to become resistant to fungicides. This will affect the crops that supply the 100 billion bananas eaten annually — the world's most popular fruit.

Most banana plantations work under contract with multi-national companies that value efficiency and low prices, raising a number of potential problems for laborers. Exposure to pesticides and other agrochemicals has been linked to respiratory problems, blindness, and sterility in Costa Rica, Nicaragua, and the Philippines. By lowering prices, supermarkets around the world suppress wages for plantation workers — men and women who work long days in extreme heat.

That simple cluster of bananas — one of many items in your cart — had a convoluted scientific upbringing, potentially injured the health of workers, required an environmentally damaging journey, and could put other banana varieties on the verge of extinction because of the decision to make an affordable, healthy, potassium-rich portable snack or breakfast cereal topping globally available.

Most foods we eat are the product of similar massive and complex systems that extend from farms and ranches to your dinner plate. These systems encompass all aspects of the food supply, a chain of events from farm production to processing, storage, distribution, marketing, retail sales, and disposal. Some food systems are relatively small-scale and local, such as products bought at farmers markets. Other food systems extend around the world and involve many intermediate steps and people, such as foods processed in factories and packaged into snacks.

Civilizations have been cultivating, processing, and cooking food for eons, and societies throughout history have been built on the back of agriculture and food systems. Today, although we're living on a populated, heavily urbanized planet, we're all still part of this ancient practice of growing, moving, selling, and preparing food. Every day, when you walk into a supermarket to stock up on staples, buy tacos from the street vendor, order groceries on your phone or stroll through a farmers market looking for that perfect tomato, you're participating in something that billions of us have shared and continue to share—our interconnected food systems—the fundamental basis of our culture, society, and survival.

The foods we eat are much more than just a source of sustenance. They have direct and substantial impacts on the nutrition and health of individuals and populations, the planet's natural resources and climate change, and structural equity and social justice challenges of societies. Food connects us to the world. It also dictates, to a degree most people don't realize, the kind of world we live in today and the kind of world we will occupy in the future.

By dramatically changing the atmosphere, the biosphere, the water cycle, and other Earth systems processes, humans have become the planet's dominant force. We've entered a new geological epoch that some have termed the Anthropocene.[1] We've built our modern world using the Earth's natural resources, but not without impacts. As the population continues to skyrocket and even more resources are required, those consequences will continue to stockpile and compound. Already, human behavior has led to global warming, habitat losses and deforestation, widespread species extinctions, and changes in the chemical composition of the atmosphere, oceans, and soil. Without drastically altering course, we'll soon struggle to feed, shelter, and treat our growing human population. Some of that behavior centers around our diets and what's on our dinner plate.

THE LINKS BETWEEN FOOD, HEALTH, EQUITY, AND THE ENVIRONMENT

Three of the biggest problems we face in the twenty-first century are (1) the burden of chronic, costly diseases such as diabetes and hypertension; (2) the consequences of climate change and natural resource degradation; and (3) the massive economic and social inequities that exist within and among nations. All three are directly related to the food we eat.

Our food systems are a wonder of the modern world. They efficiently supply almost eight billion people with enough food to survive. Annual deaths from famine fell below one million for the first time in the 2010s, and the prevalence of undernourishment has declined globally, albeit slowly in recent decades.[2] Now, many (but not all) people around the world enjoy an unprecedented quantity, quality, and variety of foods.

However, the foods we eat also contribute to increasingly common and burdensome health problems.[3] (Chapter 1 describes the connection between food and health in more detail.) Although rates of hunger have been decreasing over the past 25 years, many people still remain food insecure—not knowing when and from where their next meal will come. Many women and children still struggle with undernutrition, and obesity is rising everywhere.

More than 690 million people still go to bed hungry every night.[4] More than 2 billion people suffer from obesity, including 40 million children under the age of 5.[5] More than 20 percent of children around the world are "stunted"—too short for their age—because of a lack of nutritious foods, with most of those children living in low- and middle-income countries. At the same time, growing rates of obesity throughout the world are linked to a rise in chronic, noncommunicable diseases such as diabetes, heart disease, and cancer, which are costly, debilitating, and deadly—and are overwhelming our health systems. Without significant dietary changes, human health will further decline because of the increasing toll of these diet-related, noncommunicable diseases.

Simultaneously, food systems are placing a growing burden on the health of our planet's environment (as discussed in chapter 2). They're responsible for roughly 10 to 24 percent of global greenhouse gas emissions, which are increasing temperatures, changing precipitation patterns, and acidifying the oceans. At the same time, agricultural production is extremely sensitive to a changing climate, which will make it increasingly difficult to produce enough food for a growing population.[6] A two-way relationship exists between human activity and planetary systems. People's lifestyles and decisions are driving disastrous planetary changes, and they're also suffering from the impacts of these changes. We're victims of our own actions in a destructive feedback loop.

As chapter 3 explains, eating is an ethical act with significant implications for equity, fairness, and social justice, particularly for those who are marginalized and denied opportunities to achieve their best lives. The dietary choices of the wealthy have consequences for climate change that disproportionately affect the lives of the poor. In choosing what to eat, we're making decisions that have both short- and long-term equity implications for our global citizenry. Similarly, decisions on the efficiency and direction of food systems inevitably mean that certain moral and ethical trade-offs will have to be made. Can we sustain both human and planetary health? And, if not, what trade-offs are we willing to live with, who and what gets priority, and who and what will be left behind?

Chapter 4 discusses how we're all part of much larger social systems that will need to change to support human and planetary needs. No simple solution exists to create healthy, sustainable, and equitable diets. A constellation of different approaches and strategies — operating from the local level to global supply chains, targeting different people and organizations — will be necessary. Many solutions are available now and are ready to scale. Implementing these solutions will require individual awareness, governments' political will, and private sector investment.

Finally, in chapter 5 I discuss the actions each of us will need to take to transform food systems. Individual world citizens can help make pivotal changes through the choices they make and the policies they support. Dietary changes alone won't be enough to fix the problems, but they'll be critical in improving worldwide human and planetary health. Change often starts small and then exponentially grows. While some may think that the problem is too massive for individual action, each person has a role to play.

CAN FIXING DINNER FIX THE PLANET?

Food systems represent the nexus among diets, human nutrition and health, the environment and natural resources, animal welfare, and social equity. Changes in any of these five elements inevitably affect the other four. Our diets affect nutrition and health outcomes, even as they're shaped by consumer demands and preferences. How we grow, move, process, sell, and consume food has a huge effect on the environment and climate change, as well as on regional to global financial systems, and the laborers involved in producing and shipping the goods.

At the same time, accelerating climate change poses major risks not only to the amount of food we can grow but also to the types, safety, and quality of those foods. In turn, food systems and our diets are having a toll on finite natural resources. Left unchecked, our dietary choices will further exacerbate these problems.

Our food systems are at a critical juncture, a realization made even more plain by the COVID-19 pandemic. They have the potential to nurture human health and support environmental sustainability in equitable ways, but our current path poses immense risks. The actions we take in the next few years will set the stage for the future of food systems, as well as the future of life on this planet. If we don't address the needs of the planet, this shared ecosystem made up of humans and a vast array of other animal and plant species will struggle to survive. Your decision to put that bunch of bananas — or beef or sugar or palm oil products — in your shopping cart has a butterfly effect. It may seem like a trivial decision but it impacts the global food system, the people that shape and rely on it, and the environment that supports it.

CHAPTER 1

Are We What We Eat, or What We're Fed?

DURING MY FIVE YEARS WORKING IN TIMOR-LESTE, beginning in 2012, I saw many of the ways that food shapes the health of populations. Timor-Leste is a small island nation in Southeast Asia that's situated between Indonesia and Australia and is one of the world's youngest democracies. It has a long history of turmoil and has been ravaged by conflict and colonialism prior to its independence in 2002. About a quarter of its approximately one million people are undernourished. Of children under the age of 5, 50 percent are stunted and 38 percent are underweight. Seasonal hunger occurs each year in January and February after the previous season's harvest of rice and maize (corn, as it's commonly known in the United States) has run out and before the next harvest arrives. During this period, half the people in Timor-Leste have only enough food for one meal a day, at most.[1]

In Timor-Leste, rice is the main source of calories for most households. The country imports most of the rice that it consumes, although the national government is working to boost domestic production so that imports are no longer needed. The

Timorese people have come to view rice as a traditional food, even though it was introduced by Indonesian occupiers at the end of the twentieth century to ensure that their new territory provided their preferred food. Before that, Timor-Leste's traditional diet was comprised of roots and tubers, such as sweet potato, yam, and taro, which provide more nutritional value and diversity than a predominantly rice-based diet, and are indigenous to the Malay Archipelago. Less than half the Timorese population consumes meat and other animal-source foods regularly (usually reserved for weddings, funerals, and other special occasions), and even fewer have access to pulses (peas, beans, and other legumes) and fruits. Affordable but unhealthy junk food like instant noodles is increasingly popular and is widely available. The resulting diet has contributed to widespread micronutrient deficiencies and the overall poor nutrition of the Timorese.

Rice production has expanded to become the major agricultural investment of the government because of the country's desire to become self-sufficient, ensuring its national security and sovereignty by banking less of its food reliance on a neighbor that was recently an adversary. The rice-dominant diets of the Timorese population demonstrate what can happen when governments are concerned with ending food imports and improving food sovereignty rather than improving nutrition. This is not necessarily unjustified.

After years of conflict, the Timorese fought and won their independence. However, as my friend João Boavida, executive director of the Center of Studies for Peace and Development in Timor-Leste said, "Timor is a country that continues to live in its past."[2] Its devastating experiences with conflict inform its decision making. The pursuit of food sovereignty is based on producing enough rice to feed their population, which may not make the most economic sense. It's much cheaper to import rice from neighboring Vietnam and Thailand. By encouraging the production of more diverse and nutritious foods, including pulses, tubers, vegetables, and fruits, the Timorese government could support agriculture while building the nutritional resiliency of its people.

Timor-Leste is not alone; all countries confront trade-offs. Every decision may benefit one outcome but have negative implications for another. Timor-Leste has the right and responsibility to create its own food system to ensure its food sovereignty. At the same time, every Timorese citizen has the right to sufficient, nutritious, and diverse foods that make up a healthy diet. Diets that lack a variety of nutritious foods can impact children's physical and cognitive development, which in turn can affect their ability to earn income later. Diverse, nutrient-dense diets can also have a protective effect on adult health, helping to prevent the development of noncommunicable diseases.

If Timor-Leste chooses to accelerate its development through agriculture-led growth and rural transformation, investments in agriculture need to be substantive and diverse— moving beyond monoculture rice sovereignty. On the other hand, Timor-Leste could choose instead to grow their economy through tourism, services, and manufacturing industries. Whichever path or multiple paths it chooses, investing in the intellectual capital of their citizens will surely put the country on a path toward a knowledge economy after decades of disruption due to conflict. This means addressing the widespread malnutrition that many people in the country struggle with every day and improving cognitive development, the growth of children, and the health of future generations.

Malnutrition and food insecurity are also common in Baltimore, the city where I work. Urban poverty remains a substantive issue for Baltimore, along with other US cities like Chicago, New Orleans, and Detroit, to name a few. This poverty is often caused by entrenched racial disparities, marginalization, and discrimination. People living in these areas may have access to electricity, running water, and convenience stores (when their low-earning jobs supply enough income to pay the basic bills) but not necessarily health care, social services, and sufficiently safe-and-sound housing. Many of these neighborhoods are impacted by historic and ongoing "redlining," where cities are demarcated as being either high or low risk for investment; this discourages larger food markets from opening in the lower income areas. In places like Baltimore, these high-risk neighborhoods are typically where African American families are living.

In low-income urban neighborhoods across the city, food outlets are typically fast-food restaurants and convenience stores that offer quick, inexpensive, and generally unhealthy foods. These neighborhoods are either food deserts (where there are no markets) or food swamps (where food may be available, but shops and fast-food chains sell mainly highly processed junk foods). Not surprisingly, rates of diet-related noncommunicable diseases (heart disease, stroke, cancer, and diabetes) are much higher in these neighborhoods than in Baltimore's affluent neighborhoods that have a higher density of better quality markets with fresh, healthy foods and better health services. The inequities are stunning and obvious and sometimes exist only a couple of miles apart.

THE CONSEQUENCES OF OUR DIETS

Throughout our lives, nutrient-rich and diverse foods ideally support cognition, motor skill and social development, educational attainment, productivity, and lifetime earnings. Beyond health, foods preserve and foster social and cultural traditions that link us to other people, and meals are a prominent feature of our everyday lives. In these and many other ways, the foods and meals that make up our overall diets keep us both healthy and socially engaged and can provide the simplest of pleasures. Our days are framed and punctuated by our meals.

Dietary choices are important during the span of our lives, but they're particularly critical during periods of development such as infancy, adolescence, and pregnancy. The first one thousand days of a child's development—from conception to their second year—are pivotal for lifetime health. During this time, key nutrients such as protein, iron, zinc, vitamin A, and omega-3 fatty acids (among others) support optimal brain and immune system development and functioning, which allow children to grow and reach their full potential. But the importance of good nutrition never stops. The absence of good nutrition results in scores of physical ailments, most of which can be avoided.

Despite the significance of food for health, the quality of people's diets is diminishing worldwide.[3] In the past, most diet-related diseases and deaths were caused by caloric and nutrient deficiencies and by infectious diseases to which undernourished, poverty-stricken people were especially susceptible. Today, the causes of diet-related disease and death have shifted to noncommunicable diseases.

Unhealthy diets include those that provide sufficient energy for basic bodily functions but don't provide the nourishment to thrive. These suboptimal diets don't provide adequate vegetables, fruits, whole grains, nuts, seeds, and foods containing omega-3 fatty acids (protective fats) and contain too much red and processed meats (salted and cured), such as salami and hotdogs. There's also a high intake of salt, unhealthy fats, and sugar bundled up into highly processed, packaged foods that Michael Pollan, the journalist who wrote *The Omnivore's Dilemma*, called "edible foodlike substances." Overconsumption of sugar-sweetened beverages such as soda and sweetened ice teas is prevalent in every country.[4] The diets we eat have now become the number one risk factor for preventable deaths, contributing to lifelong disabilities and nearly 11 million deaths in 2017.

Ashkan Afshin, a colleague of mine who's a professor at the University of Washington, leads the dietary work for the Bill and Melinda Gates Foundation–funded Global Burden of Disease project, a collaboration of more than 3,600 researchers representing 145 countries. He offered some insight on what diets will look like in the future: "Recently, we forecasted the burden of disease related to various risk factors over the next three decades under different scenarios. Our results showed that under almost all scenarios, diet and diet-related risk factors (i.e., obesity, high blood pressure, and high fasting plasma glucose) were among the top five risk factors for mortality globally."[5]

How have we gotten ourselves into this ironic situation in which diets meant to nurture us are essentially killing us? There's no easy solution, and there's no single culprit. Many factors are at play, including people's food choices shaped by personal preferences and situations, food environments, global trade and food supplies, and social and cultural factors, such as education, income, employment, and housing. All these factors intersect in complex and shifting ways to determine what and how people eat. They also shape the vast inequities we see across diets and explain why some people remain healthy while others develop diet-related diseases.

Are Processed Foods Bad for You?

Highly processed foods can be nutrition landmines, as they generally include unhealthy levels of fats, salt, and sugar. Some of these foods also prompt environmental concerns because of the energy-intensive production required to create specific products, the fossil fuels they burn through transport stops along their delivery chain, their dumped by-products, and the landfill-jamming packaging that houses them. But what does the term *processed* really cover, and is it always a bad thing?

Essentially, a processed food has been altered from its natural state to prepare it for consumption, to extend its shelf life, or both. Most food is processed in some way, except for items including eggs, fresh fish, and produce from a local organic farm. It's a broad spectrum of actions, most of which cause no physical harm to consumers in moderation. Processing can be as simple as dyeing oranges and other fruits and vegetables to make them more appealing and pressing olives to make oil, or as complex as creating junk food such as cheese puffs or frozen pizza with toppings that have each gone through many processing steps. It includes ancient processes such as milling (which can strip out fiber, B vitamins, and some minerals from the grains while making them more digestible); preservation techniques such as canning, freezing, and fermenting (which reduce foods' nutrients through exposure to high levels of heat, light, and/or oxygen but can also improve their probiotic potential); and modern technological additives that introduce petroleum-derived products (chemical food dye, mineral oil, paraffin wax) into foods.

"Processed" in and of itself isn't bad—it's the *highly* processed, overengineered foods that should be limited, as they've been associated with dangerous effects on health, such as heart disease and weight gain. These foods trigger our brains' pleasure centers (delicious! crunchy!) and satisfy addiction-like cravings. They're fine as occasional treats, but if their nutrition labels include ingredients that you know are unhealthy or have never heard of, you might want to reconsider. Also, remember that while the US Food and Drug Administration regulates both natural and chemical ingredients, the FDA bases safety on certain serving sizes, so higher volume ingestion has not necessarily been proven safe. When you're choosing foods, consider:

- how many steps they've gone through to get to the store shelves in their final forms and how you're going to use them (if the milk you bought is for a cream sauce that's then baked, you aren't going to get the same nutritional benefit as if you drank a cold glass of it);
- the origins of those ingredients, and what harm to the planet and its residents may occur in generating them (such as the destruction of orangutan habitats when palm trees are clear-cut to process palm oil);
- how many unrecognizable additives it has; and
- how far it traveled.

The fewer degrees of separation from the soil or water to your plate, the better for your waistline, heart, and the planet.

FOOD SECURITY

Food security is a major factor in whether a person suffers from malnutrition. People are food secure when they have reliable physical access to a sufficient and stable quantity of affordable, nutritious, safe, and diverse foods. Governments, agribusiness, and regional and international trade organizations must do their parts to ensure that enough food is produced globally, that the food produced is sufficiently diverse, and that it's effectively moved to markets without significant loss.

Even when food is available, accessible, and affordable, the human body also needs to be able to use it. Effective food utilization depends on household knowledge about safe and healthy food storage and preparation techniques, food waste, and the efficiency with which people absorb and metabolize nutrients. Individuals with frequent or chronic infections may have impaired absorption of nutrients, hindering their ability to efficiently use the foods they consume.

When I traveled to southwestern Uganda on multiple visits between 2007 and 2010 to assist agriculture and nutrition programs in local communities, food insecurity and child malnutrition were stunningly high. This was perplexing because this region is the country's breadbasket, supplying the nation's calories. Subsistence farmers grow matoke — plantains (a starchy relative of the banana) — used for their national dish of the same name, cooked with spices. Why is poverty high and dietary diversity low in such a fertile, productive area? One reason is that the way plantains are raised leaves little room to grow other foods. Farmers clear-cut the land, chopping down other trees along with all the stumps and brush, to make room for the plantain trees. In the absence of electricity families rely on fires to cook their food, but deforestation leaves little firewood or fuel, particularly for food that takes longer to cook, such as healthy beans.

Farmers certainly won't cut down their major cash crop, the plantain, to prepare meals. Therefore, 70 to 80 percent of a southwestern Ugandan family's diet consists of matoke, which is calorically rich but nutritionally poor. Farming families are incredibly vulnerable if an infestation or frequently occurring climate-related natural disasters destroy their crop. In addition, malaria, diarrhea, and communicable diseases hit these communities hard, leaving people (particularly children) too sick to absorb what nutrients remain in their monotonous diets.

My friend and colleague Enock Musinguzi, who is from southwestern Uganda and works at the Global Alliance for Improved Nutrition (GAIN), said, "Africa has some of the richest areas in terms of agrobiodiversity and hence are the food baskets for their countries. But many of these same areas are teeming with the most severe and debilitating forms of malnutrition. From the southwestern part of Uganda to western Kenya through the southern agricultural corridor of Tanzania to mention a few, this 'scarcity amidst plenty' continues to rear its head and manifest itself in an uncomfortable but familiar fashion."[6]

Poverty both causes and results in food insecurity. More than one billion people around the world continue to live in extreme poverty, earning less than $1.25 a day, and this number dramatically increased with the COVID-19 pandemic that sent unemployment numbers skyrocketing and businesses into bankruptcy. However, economic growth and stability alone aren't sufficient to eliminate food insecurity and hunger. Inequalities exist between urban and rural areas in many countries, across regions or between various ethnic groups. For example, India has had enormous economic growth in the last decades but still faces significant burdens of child and maternal undernutrition and debilitating diabetes and obesity trends.

More than two-thirds of those living in extreme poverty go hungry, and children in these families are five times more likely to die before age 5.[7] In addition, malnutrition harms a person's ability to earn a living, creating a vicious cycle between poverty and malnutrition. Food insecurity and poor diets cause increased susceptibility to many communicable and noncommunicable diseases, reduced capacity for physical work, lowered cognitive capacity, increased exposure and vulnerability to lifestyle-related and environmental risks, reduced participation in social decisions, and increased difficulty handling environmental challenges. It's a poverty trap that's nearly impossible to escape.

Chronic food insecurity at the household, community, or societal level can lead to the "double burden"—both undernutrition and obesity. People who are food insecure tend to eat either subsistence staple foods (such as those mentioned in southwestern Uganda or Timor-Leste) or inexpensive, high-calorie, low-nutrition foods (as in Baltimore). Although food-insecure populations may get access to nutritious diets or foods once in a while, these periods tend to be punctuated by cycles of financial and personal stress. The result is often food deprivation or overeating less healthy foods, limited access to health care, reduced opportunities for physical activity, and greater exposure to unhealthy or insufficient food environments. Empty calorie diets that either lack variety or rely on highly processed foods can cause weight gain without providing nourishment. And a significant body of evidence suggests that hunger in utero and in early life can put individuals at higher risk for becoming overweight in adulthood.

These observations apply around the world. In the United States, poverty and unemployment have driven the dual rise in food insecurity and obesity since the 1960s, especially in rural areas. People who live in cities also are susceptible to food insecurity, as many urban settings lack adequate services and support structures to ensure nutritious diets. Nearly 12 percent of American households are food insecure, which is staggering considering this nation's wealth. This amounts to roughly 40 million people, including roughly 540,000 children who experience very low food security.[8] Food insecurity tends to be highest among Hispanic and Black non-Hispanic families, and among the unemployed, households with children headed by single women, and the poor across all communities. Even in high-income countries, inequities can be profound. They exist in what are perceived as wealthy small communities, such as resort areas where older generations on fixed incomes and service workers live among or near wealthy second-home owners and vacationers. Often those who've lived in these areas for decades can no longer handle their rising property taxes as their neighbors build mansions. Budgets for food and other necessities are cut short in an effort to keep their homes.

It may sound contradictory, but many people who go to bed hungry also struggle with weight gain. Research has found that food insecure adults in the United States are 32 percent more likely than others to be obese, especially if they're women.[9] Another study discovered that children living in food insecure households have a greater-than-average tendency to be overweight or obese, along with having poor eating habits.[10] Moreover, food insecure children tend to exhibit significant behavioral problems, disrupted social interactions, poor cognitive development, and marginal school performance, all of which increase their risk of becoming obese adults who will face difficulty in finding and keeping jobs.[11]

TRANSITIONING DIETS

Our diets are changing and have been for decades. The overall trend is that we're eating more calories, and global diets are less nutritious, environmentally unsustainable, and inequitable. Global food supply data show that in 1960 the world consumed roughly 2,200 calories per capita per day; now, it's 2,800 calories.[12] People in America consume on average 3,600 calories per day (recommended intake should be around 2,100 calories per day). Much of the caloric climb has to do with the doubling of portion sizes over the past 20 years.

We're not only eating too many calories, but we're getting them from substandard sources. We're not eating enough of the good stuff—fresh fruits and vegetables, nuts and seeds, legumes, and whole grains.[13] Instead, the trend is to consume more processed animal products, oils, sugar-sweetened beverages, and highly processed packaged foods laden with added sugars, unhealthy fats, and salt. What are examples of these highly processed "junk" foods? Think cellophane-wrapped cakes, instant noodles, candy, and chicken nuggets. These foods permeate stores all over the world but are particularly on the rise in Asia and Latin America.[14]

There's also an increasing demand in many parts of the world for certain foods that tax the environment.[15] The process of raising animal-source foods such as beef, lamb, farmed shrimp, and cheese emit high levels of carbon. Cheese, nuts, and farmed shrimp use a ton of water to produce 1 kilogram of product. Many animal foods such as beef, dairy products, pork, and poultry, cause significant nutrient runoff (also known as eutrophication) into waterways and ecosystems from their waste, causing dangerous algae blooms and other problems that kill marine life. The global consumer demand for meat is rising from 20 kilograms per capita per year in the 1960s to 45 kilograms in the present time. Wealth is one of the strongest determinants of how much meat people eat. Australia, Europe, and North America remain the highest consumers of meat, averaging somewhere between 100 and 115 kilograms per person per year. As a stark contrast, India, a largely lacto-ovo vegetarian (milk, eggs, and vegetables) population, consumes only 4 kilograms of meat per person per year, and in Ethiopia it's less than 10 kilograms.

These trends in animal food consumption highlight inequities in who gets access to these more "luxurious" and costly foods. As incomes go up, and with urbanization, people have a greater demand for their food supply to deliver more diversity. There's a demand not only for more types of foods but also for these foods to be prepared in different ways. As people become wealthier they have more options to eat either nutritiously or, if they choose, unhealthily. Those who remain poor have limited options and often can only afford the cheaper calories from staple grain products or shelf-stable, processed foods. For some low-income households, a "healthy" diet is unfeasible. Organic food, farmers markets, and specialty products are out of reach both literally and financially.

A few factors shape these dietary transitions, including globalization, trade, urbanization, and rising incomes. Trade and globalization have allowed for food supply chains to move more food products around the world, including to remote places, as infrastructure networks develop and agreements between nations evolve. Mexico has undergone a tremendous transition over the past 30 years, moving away from its traditional diets of maize and beans to diets that mirror the United States; Mexican consumption of soda is now the highest in the world. The changing food environments of Mexico can be partially blamed (not completely, given that Canada has not experienced the same transition) on the North American Free Trade Agreement, or NAFTA, which allowed free-flowing trade goods to pass from the United States to Mexico, speeding up their dietary transition and obesity prevalence.[16]

The Brazilian Amazon, a mosaic of thriving indigenous communities, has significantly changed in the past decade. The infiltration of multinational companies through trade agreements and globalization along the Amazon riverways has been a driving source in changing the traditional diet, reliant on the rich biodiversity of the Amazon, to mimic the worst traits of an American diet.

Changes in income and routines affect food preferences. As people have more disposable income, they can make other purchases besides food. They spend their hard-earned money on a diverse set of foods prepared in different ways. In Nepal, the traditional *dal bhat* (a meal made up of steamed rice and lentil soup) takes hours for women to prepare in the morning. That more nutritious meal is being replaced by instant noodles. They are cheap, take three minutes to prepare, children love their taste, and it saves moms hours of time in the kitchen.

Increasing urbanization also tends to produce lifestyle changes, new job opportunities wherein both parents work, and a greater reliance on services within an economy. People cook less, eat out more, and shop more at larger supermarkets than in local grocery stores. In 1990, 10 to 20 percent of Latin Americans shopped at supermarkets, while the majority shopped at smaller local markets. Over the course of one decade, with significant urbanization in many countries, that number has risen 50 to 60 percent. Compounding the impact of their food selections on health, people increasingly have sedentary jobs in cities and tend to be less physically active overall than in rural areas. They rely on public transport and cars, walk and bike less, and burn fewer calories. China, for example, was a nation of bikers. Now, with massive urban growth, it's a nation of car drivers or public transport riders. Higher incomes in China are also associated with increases in caloric intake.

These factors — growing cities, increasing incomes, increasing globalization — are changing people's diets and their nutrition outcomes. This change away from traditional diets consisting of indigenous, and often (but not always) healthy foods to modern diets and sedentary lifestyles was described as the "nutrition transition" by economists Barry Popkin and Adam Drewnowski in the early 1990s. This transition shows that as countries industrialize and diets and lifestyles change, people suffer less from hunger and food insecurity and begin to struggle with obesity and diet-related noncommunicable diseases. This trend has been witnessed in every country in the world. China has seen massive reductions in hunger and undernutrition over the past 25 years, but diabetes and strokes are rising. Increasing urbanization is correlated with increased incidence of diabetes, obesity, and high blood pressure.

For two centuries, many high-income countries have experienced the nutrition transition at a very slow, gradual pace. But in many low- and middle-income countries, these processes have been occurring in decades rather than in centuries. In most countries in the world and in places undergoing rapid food systems transitions, overweight (a body mass index, or BMI, over 25), obesity (a BMI over 30), and noncommunicable diseases are gradually replacing the health problems caused by undernutrition. As a result, people live longer, but they have higher disability and thus suboptimal qualities of life. What countries are in this phase? India, China, Thailand, Egypt, the Middle East, South Africa, and Mexico, to name a few. As low- and middle-income countries gain wealth, they have opportunities to avoid the negative dietary trajectories that other countries have created. But doing so will take a conscious and concerted effort geared toward large-scale food system transformations.

UNDERNOURISHMENT AND INADEQUATE DIETS

Undernourishment is a biological consequence of food insecurity. It occurs when a person doesn't ingest enough protein, fats, calories, or micronutrients for their growth and health. Those particularly vulnerable are unborn children, infants and children under the age of 5, women of reproductive age, and older persons. Undernourishment triggers lifelong and intergenerational consequences. It has harmful effects on a child's physical and cognitive development, reduces quality of life throughout life spans, and decreases resistance to infection in people of all ages.

Undernourishment often results from the insufficient and unbalanced diets typical of those experiencing food insecurity. Even if lower- and middle-income populations eat more environmentally sustainable diets of plant-based foods, these diets can sometimes lack the full range of nutrients required to support good health if the balance and composition of the meal is not sufficient. Many of these populations subsist on grain- or tuber-based diets (for example, maize, rice, wheat, potatoes), which can be harder to digest if not processed correctly and don't have the same amino acid profiles as animal-source foods and high-protein, plant-based foods such as legumes and nuts. Grain-dominant diets also lack key micronutrients that are important for growth, immunity, and cognitive functioning and can also have compounds that reduce the absorption of some micronutrients (such as phytates and oxalates).

Populations living in starkly poor areas of the world are often vulnerable to infectious diseases, "wasting" (acute undernutrition) or "stunting" (growth impacted by chronic undernutrition), high maternal and child mortality, and other health problems.

People's homes often don't have electricity or running water and food is frequently cooked over fires. These conditions, in addition to the inability to afford a wide array of foods, limit the kinds and amounts of items they buy to feed their families, which in turn makes them susceptible to those serious health concerns. These open fires or simple stoves using biogas, kerosene or coal generate indoor air pollution and expose households to health-damaging pollutants that can lead to pneumonia, chronic obstructive pulmonary disease, and other health issues.

When I worked in the rural parts of Malawi and Kenya, I noticed that people's plates, for all meals, are mainly filled with maize, the staple grain. In Rwanda, it's cassava, the carb-laden, tapioca-producing root of the tree of the same name, which, while containing some vitamins and minerals (and protein, if the leaves are eaten), doesn't provide all a body needs. In Senegal, Myanmar, and Cambodia, it's rice. None of these predominantly single-item diets contribute to healthy communities.

In addition to poor diets, access to clean water plays an important role in nourishment. Approximately 844 million people worldwide, or more than 1 in 10, don't have access to clean water.[17] Without such access, people are forced to drink unsafe water from unprotected sources. Each year, millions of people living in the Least Developed Countries (as identified by the United Nations) die from diseases introduced by tainted drinking water and poor sanitation. These waterborne diseases strip out what little nutrients people have in their systems and make the sick unable to eat through the course of the ailment, further deteriorating their nutrition and immune status. Diarrhea is one of the leading causes of death among children around the world.

Contaminated food is another major source of undernourishment morbidity and mortality in low- and middle-income settings. Aflatoxin — a poisonous by-product of the Aspergillus fungus found in warm, humid areas of the world — can cause significant damage to the liver. It commonly contaminates maize, peanuts, and tree nuts that aren't properly dried and stored. These contaminants are associated with growth impairments in children, are dangerous carcinogens at low levels, and can be fatal in high doses. The food supply is the source of many foodborne illnesses and we're now finding that zoonotic diseases (viruses passing from animals to humans) such as COVID-19 are also thought to originate through food. Other toxins, such as pesticides and herbicides used on crops, have unknown long-term health effects, particularly on children.

CHILDHOOD UNDERNUTRITION

The most frequent result of chronic undernutrition is stunting — preventing children from reaching their potential heights for a particular age. A stunted 5-year-old is on average 4 to 6 inches shorter than their nonstunted classmates. But the problem goes much deeper than short stature (full disclosure, I'm 5′1″, well below the average height!). These children are not only shorter than their peers, their brains don't grow or function as well as those children who have good nutrition. When over half the population of a country is unable to reach their full height capacity, the implications for economic development are severe.

When I worked in places with high burdens of stunting, such as Bangladesh, Ethiopia, and Nepal, it was really hard to tell which child was stunted and which was not. It's not something that's obvious to the eye, which makes it so hard to raise awareness about this public health problem unlike, say, Ebola, which is visceral and shocking to witness. However, stunting affects approximately 140 million children under the age of 5 worldwide, with the majority of cases in Africa and Asia, where prevalence can exceed 30 percent.[18]

Crucial determinants of stunting include poor maternal health and nutrition before and during pregnancy and lactation, inadequate breastfeeding, inadequate maternal diets that compromise breastmilk quality, poor feeding practices for infants and young children, and unhealthy environments (including a lack of access to clean drinking water, sanitation, and hygiene) for children. Stunting often begins in utero and the crucial early years of development. Political, economic, environmental, and cultural factors all contribute to stunting.

Stunting has both short-term and long-term consequences. In the short term, stunting increases the risk of mortality, though it's not a primary cause of mortality. In the long term, stunting affects the health, education, and productivity of children as they become adults. In addition to height deficits, stunting is associated with (though not the direct cause of) cognitive disability, decreased motor skills, and compromised immune function. These effects can lead to reduced adult income and an increased risk of developing noncommunicable chronic diseases. If and when the body finally does receive the nourishment it requires, the resulting weight gain can increase the risk of adult obesity. Some young children are both stunted and overweight. This double burden is found in rural Mexico and India for example and is strongly tied to younger mothers, lower socioeconomic status, lower education level of mothers, lower stature, and household size.[19]

Another type of undernutrition is acute malnutrition, which is often manifested by "wasting." Wasting commonly results from temporary or cyclical events such as natural disasters, conflicts, seasonal depressions, or highly infectious disease environments. In food-insecure communities in Bangladesh, the monsoon season can cause decreased dietary diversity and lower household income, which often leads to wasting.[20] Wasting is the result of a body's rapid consumption of its fat, muscle, and tissue supplies until there is no longer any fat and tissue left to consume, after which growth is interrupted. Wasting affects approximately 10 percent of children under the age of 5, roughly 50 million.[21] Although more children are stunted than wasted worldwide, more children are likely to die from wasting than from stunting.

The effects of undernutrition transcend generations. India, Nepal, and Bangladesh are often caught in this cycle. If a woman enters pregnancy with poor nutritional status, she's more likely to have a low-birth-weight infant with an increased risk of developing chronic diseases in adulthood. If a child doesn't receive the right nutrition and care early in life, then that child is put on an unfortunate path. Stunted children become stunted adolescents who have difficulties in completing their education, leaving them with fewer opportunities to earn living wages. Girls become stunted adults who potentially give birth to stunted infants. Furthermore, the earlier a child undergoes stunting, the greater the long-term effects.[22] It's essentially a life sentence.

Undernutrition is like a snapshot of a country's history. The high stunting burden in Timor-Leste reflects a long history of conflict, colonization, and occupation. Portugal controlled Timor-Leste for centuries, until the resistance movement won independence in late 1975. Indonesia then quickly invaded and occupied the country. The Timorese fought bravely until they won their independence again, in 1999, though at least 25 percent of the population was killed in the process. In the years since 2002, when the country became fully independent, several smaller conflicts have flared up, requiring interventions by United Nations peacekeepers. Only in the past several years has Timor-Leste been peaceful and stable enough for its leaders to focus on improving people's lives and diets. As Timor-Leste adapts to this new, conflict-free state, it's hoped that stunting will become less prevalent.

Other countries with the highest burdens of stunting and wasting—Afghanistan, Yemen, Madagascar, Nepal, Mozambique, Somalia, and the Democratic Republic of Congo—are similarly either rife with conflict, whether within their own nation or between nations—or recovering from conflict. But conflict doesn't inevitably result in undernutrition. Despite its low-income status as a postconflict nation, Nepal's population has experienced remarkable progress in reducing stunting over the past decade.

My colleague at Johns Hopkins Dr. Swetha Manohar says that Nepal has seen significant declines in stunting because:

> the end of the civil war in 2006 brought a lot of investment in rural development with roads being built, and health facilities and schools being rebuilt, allowing for greater community access to them. There was also a large push to improve sanitation—to build toilets for homes. But you can't talk of this development without speaking of the out-migration of an enormous number of Nepalis who sought work elsewhere but contributed to increasing their household wealth and the country's GDP [gross domestic product] by sending home remittance money. This had a profound impact on families' income to spend on education, health, diversifying their livelihoods, and improving their standard of living.[23]

While all these improvements contributed in direct and indirect ways to reducing stunting in Nepal, the progress was unequally experienced between the children of the rich versus the poor, the urban versus the rural, the majority versus the minority ethnic and caste groups, and the children of formally educated versus uneducated families. The latter of each of these groups saw slower rates of decline in stunting and experienced higher burdens of stunting compared to their more privileged counterparts.

OBESITY

Another type of malnutrition is obesity. Obesity plagues nearly every country on the planet and has risen all over the world, driven largely by global dietary shifts, sedentary lifestyles, and urbanization. In the United States and the United Kingdom, the rate of adult obesity has tripled over the past three decades; more than one-third of adults in the United States are obese. Worldwide, 2.1 billion adults are overweight, and 678 million of these adults are obese, compared with 462 million adults who are underweight.[24] In 2000, 30 million children under age 5 were overweight; today, approximately 38 million children under age 5 are overweight, and two-thirds of these children reside in low- to middle-income countries.

This epidemic of obesity is taking a heavy toll on human health. Overweight children are at increased risk of developing heart disease and type 2 diabetes later in life. The prevalence of diabetes among adults increased 81 percent from 1980 to 2014, swelling to an estimated 422 million cases.[25] Globally, obesity costs an estimated $2 trillion annually, or 2.8 percent of the global GDP, driven by health care expenditures, lower productivity, mortality, and permanent disability.[26]

Overconsumption of food is not always the result of conscious food choices. Home, work, and school environments increasingly lead people to overeat and eat poorly. Particularly in middle- and high-income countries, disadvantaged, vulnerable, and poor people often find themselves in food environments that promote obesity, but this is quickly becoming the norm in low-income countries as well. Unhealthy triggers may be automatic, such as getting fries with a meal. These responses are often prompted by cues or "nudges" in the immediate environment, such as where food is placed in the grocery store (candy at the checkout line); tray and plate sizes in cafeterias; buy one, get one free deals; and so on. Continual repetition and reinforcement of these behaviors can make it extremely difficult to break unhealthy habits and lose weight.

Food systems now provide nearly universal access to unhealthy and highly processed foods. Even outside the traditional food system—in hardware stores, gas stations, and bookstores—food is sold, especially long shelf-life, processed foods that people buy impulsively, not necessarily when they're hungry. The portion sizes of many packaged, restaurant, and takeout foods have increased while their relative costs have decreased. At the same time, the cost of fresh produce has increased, particularly among poor people in low- and middle-income countries and in countries that rely on importing fresh food. Industrialized, modern food systems place an increasing burden on consumers to make the right food choices amid a dazzling display of diversity, despite insufficient and often contradictory information on health and nutrition.

Malnutrition can alter gut microbiomes in such a way as to increase the risk of disease. Microbiomes are communities of trillions of diverse microbes that consist of fungi, bacteria, and viruses. In healthy populations, these microbes are harmless and live in symbiosis with humans. They live in various parts of our body including our intestinal tract, behind our ears, and under our armpits. Each person has a unique microbiome; our exposure starts in the womb. Environmental factors and diets can change our microbiome, promoting or harming health depending on how much we disturb its balance. Disruption of the gut microbiota during childhood can lead to various metabolic syndromes, including obesity. Studies in Malawi and Bangladesh have shown that children's compromised microbiome composition contributes to wasting. Certain foods, known as prebiotics (non-digestible plant fibers found in foods) and probiotics (live bacteria cultures found in foods), can "nurture" the microbiome to aid in immunity and regulate blood glucose levels and cholesterol.[27]

In Myanmar, I worked with my former postdoc and now colleague Shauna Downs, a professor at Rutgers University who does innovative work on food environments. We were trying to understand how diets are changing in both rural and urban places and how these changes are contributing to obesity. After enduring 60 years of conflict, Myanmar has been on a path to open, sustainable development since 2010. But along with its recent economic growth has come an increase in both under-nutrition and obesity. Obesity afflicts 30 percent of women in Myanmar, while 29 percent of children are stunted. The diets of the Burmese are diverse but low quality. Many Burmese say that they prefer fruits, vegetables, and red meat to highly processed snack foods and beverages, but they tend to consume high quantities of high-fat, high-sodium Burmese street food because it's inexpensive, easily found, and satisfying.[28] When Shauna and I ate street food during our visits we often found an inch or two of oil on the bottom of our soup bowls after gobbling up the delicious curries. Many Burmese talk about how oily the street food is now, a big change over the past 20 years. They also remark on the opening of the first fast-food restaurant in Yangon, Kentucky Fried Chicken. The queues were long, the crowd was youthful, and the selfies were abundant.

THE DOUBLE BURDEN OF OBESITY AND UNDERNUTRITION

Counterintuitively, undernutrition can coexist with obesity. Individuals can suffer from both types of malnutrition simultaneously or they can develop multiple types over a lifetime. Households can have multiple family members affected by different forms of malnutrition, as can communities, regions, and nations.[29]

Undernutrition and obesity have a biological and physiological connection. A mother who's undernourished or obese, and the environment and diets she's exposed to, can have ramifications on the fetus, creating genetic and physiological changes in children that can manifest into adult life. Some of this connection is due to the intergenerational cycle of undernutrition stemming from energy-inadequate diets. If a mom is undernourished in early life, this will adversely affect early growth and development of her fetus, likely leading to a stunted child and teenager. Early stunting predisposes an individual to a higher risk of obesity at later ages.

An overweight mom who is exposed to unhealthy diets along with other environmental triggers such as psychosocial stress, insufficient sleep, and physical inactivity can put her child at risk of early onset obesity. Layer on the nutrition transition, with countries and communities at different stages of changing economies, globalization, and urbanization, and we're left with populations still struggling with both undernutrition and emerging obesity.

The double burden of malnutrition and obesity has increased rapidly in recent decades and is causing higher morbidity, mortality, disability, and health care costs. It creates a vicious cycle in which undernutrition predisposes a person to obesity and obesity contributes to undernutrition. Combined, these effects can perpetuate a cycle of poverty and poor health.

MICRONUTRIENT DEFICIENCIES
————

Inadequate consumption of essential vitamins and minerals can result in micronutrient deficiencies known as the "hidden hunger." All ages and populations are affected by such deficiencies, which can disrupt many biological functions. Consequences of micronutrient deficiencies can range from nonspecific effects such as impaired cognition, growth, or immune functions to specific conditions such as scurvy (vitamin C).

At least half of children worldwide ages 6 months to 5 years suffer from one or more micronutrient deficiencies. Globally, based on national survey estimates, more than two billion people of all ages are affected.[30] Inadequate diets, poor sanitation and hygiene, and broken public health systems all contribute to significant and multiple micronutrient deficiencies.

The World Health Organization has classified 153 countries as having an anemia prevalence that has moderate to severe public health significance.[31] One in three women of reproductive age suffers from anemia, a condition characterized by abnormally low blood hemoglobin concentration, which can have lifelong effects.[32] Maternal anemia increases the risk of maternal mortality, preterm delivery, low birth weight, fetal malformations and deaths. Anemia can be caused by iron deficiency but also other issues such as blood disorders and heavy menstruation, to name a few. Iron is essential in the body for oxygen transport and cellular respiration, without which the brain and muscle cells cannot work. Iron deficiency in children is associated with poorer cognitive outcomes and school performance.

Other common micronutrient deficiencies include zinc, vitamin A, iodine, folate, vitamin B12, and other B vitamins. Getting too little vitamin A can cause dry eyes, night blindness, an increased risk of infection, and child and maternal mortality. Symptoms of zinc deficiency include loss of appetite, stunted growth, delayed healing of wounds, hair loss, and diarrhea. Iodine deficiency can cause enlarged thyroid glands (goiters), decreased production of thyroid hormone (which affects metabolism), and growth and development problems. The severe effects of micronutrient deficiencies highlight the importance of not only eating a healthy diet but also eating a varied diet that meets the body's micronutrient needs.

SOCIAL NORMS AND CULTURAL TRADITIONS

While diets and food system inequities are most certainly shaped by a variety of global and national forces and actors, diets are also shaped by societal customs and beliefs that reflect the knowledge, traditions, and norms of people and cultures.

Cuisine types, food traditions, and cultural dietary customs can represent a person's heritage. Mediterranean, Nordic, and Japanese diets are shaped by their locale and traditions. These regional diets are part of larger, territorial food systems. As such, the familiar foods consumed regularly in these regions can be powerful in preserving social traditions and norms. Religion can also shape the values and practices that people bring to food consumption, such as those who practice halal (Muslim) or kosher (Jewish) strictures.

In my Italian American (and Catholic) family, we never consumed meat on Fridays during Lent—the staple for dinner on those Friday nights was always fish. And when I was working in northern Ethiopia from 2008 to 2012, I followed the modified form of fasting of the Ethiopian Orthodox Tewahedo Church. It required abstaining from meat, dairy, and eggs once or twice a week—and on other longer occasions throughout the year. The types of foods we consume, the ways in which we prepare and cook meals, and the ways in which those meals are consumed form the foundation of many traditions, religions, and cultures worldwide.

Special occasions and celebrations may call for particular foods. In many parts of sub-Saharan Africa—despite widespread rural poverty—communities contribute substantial time and resources to honoring deaths and marriages. When possible, these events typically occur when people are able to contribute resources to purchase higher-quality foods such as animal-source foods. These traditions continue and strongly reinforce social traditions.

Dietary restrictions, taboos, and rules may also relate to particular phases of the life span, such as infancy and childhood. Pregnant and lactating women are often told to avoid caffeine, alcohol, and fish, mainly for health reasons. In some parts of Western and Eastern Africa, women are told not to eat eggs because it could make them sterile.

Food systems will face major challenges in the years ahead. One of the primary goals of food systems is to supply healthier foods to the world's population to help prevent malnutrition and ensure that everyone is able to access a healthy diet. If we continue on the current course with multiple burdens of malnutrition affecting every country in the world in some way, shape or form, it will be very difficult to "fix" planetary health. Without healthy humans, there can't be a healthy planet, and with poor planet health there will be poor human health. They go hand in hand. The cords that bind them are our diets and the food systems from which those diets originate.

But to fix planetary health, as chapter 2 explains, we'll have to deal with an ever-increasing threat to food systems: climate change. Two prominent public health specialists, Boyd Swinburn and Bill Dietz, call the intersection of undernutrition, obesity, and climate change a "syndemic"[33]—two or more pandemics that are tethered in time and place and are driven by societal and economic factors. Boyd called this syndemic the paramount health challenge for humans, the environment, and our planet in the twenty-first century.

CHAPTER 2

Can Cooking Curry in Cambodia Trigger a Tornado in Texas?

IN 2018, I SPENT A YEAR IN ROME working at the Food and Agriculture Organization (FAO), the leading institution on food systems and agriculture of the United Nations and the agency charged with tracking and addressing global hunger. After 15 years of declining rates of hunger around the globe, 2017 was the first year that the rates increased. During my time at the FAO, I had many discussions with colleagues about why the formerly favorable trend had reversed. Gradually, we realized that two major factors, conflict and climate change, were responsible for the increases in hunger from year to year.

It was an unprecedented time. After making great progress, we were starting to see a regression. It was a blow to realize that climate change, conflict, and the resulting food price shocks were leaving more people vulnerable to hunger and poor health outcomes—and that those adverse outcomes could continue to increase. Even more devastating was the realization that such outcomes could keep reoccurring over and over again in the future.

A BIDIRECTIONAL RELATIONSHIP

Climate change affects every aspect of food systems. Without action, it's expected to cause a 2 percent decrease in food production every decade until 2050, and much more drastic decreases after that.[1] At the same time, practices within food systems essentially affect all environmental systems. Those food systems (and, by extension, our diets) and the environment are thoroughly intertwined.

The Anthropocene is the current geological period we're living in, in which humans have become the dominant influence on the global warming of the planet, rising sea levels, animal and plant extinctions, and habitat loss. Agriculture has been a major contributor to the environmental changes of the Anthropocene, now using 37 percent of the Earth's land and 70 percent of its freshwater supply.[2] It's the biggest source of nutrient runoff, causing algal blooms, dead zones, and acidification of the planet's freshwater and ocean ecosystems. These changes, along with the accelerated clearing of forests for agricultural use, have contributed to one of the significant events of the Anthropocene Epoch: a mass extinction event where, since 1970, there has been a decline of 60 percent, on average, in the number of species on Earth for mammals, birds, fish, reptiles, and amphibians. Currently, agriculture (including raising animals) contributes between 11 and 24 percent of total greenhouse gas emissions.

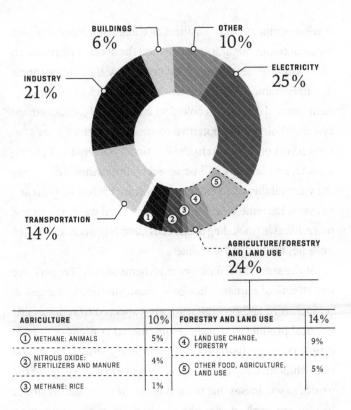

AGRICULTURE	10%	FORESTRY AND LAND USE	14%
① METHANE: ANIMALS	5%	④ LAND USE CHANGE, FORESTRY	9%
② NITROUS OXIDE: FERTILIZERS AND MANURE	4%	⑤ OTHER FOOD, AGRICULTURE, LAND USE	5%
③ METHANE: RICE	1%		

GLOBAL EMISSIONS BY SECTOR: The agriculture sector combined with forestry and land use accounts for 24 percent of global greenhouse gas emissions (GHGe), including some of the more toxic gases such as methane and nitrous oxide. Methane from grazing animals and rice and nitrous oxide from fertilizers and manure make up 10 percent of those GHGe.

J. Poore and T. Nemecek, "Reducing Food's Environmental Impacts through Producers and Consumers," *Science* 360, no. 6392 (2018): 987–992; and *Our World in Data*, https://ourworldindata.org/environmental-impacts-of-food.

If we continue on this "business as usual" trajectory, in which we compromise natural resources and the overall environment, the consequences for our food systems will be catastrophic and irreparable. Continued deforestation to clear land for agriculture will cause the ecological collapse of biogeochemical systems. This has the potential to affect the entire planet's oxygen levels on land and in the oceans. Biodiversity of plants, insects, and animals will be severely diminished, increasing the vulnerability of ecosystems important to food systems and humans. Extreme weather events, food and water shortages, more diseases (including pandemics), and other climate-related maladies are likely to skyrocket.

At the same time, food systems themselves will experience the effects of climate change. Climate-induced changes in temperature and precipitation are expected to reduce agricultural productivity and the nutritional content of certain crops, compromising food availability, consumption, and nourishment. Crop diseases will increase in some parts of the world, as will losses due to pathogens infecting grains stored in silos and sacks during the postharvest stage. Combatting these outbreaks will require more and better cold storage of food, which in turn will use more energy and adversely affect climate change unless we move aggressively to renewable energy resources. In the years ahead, food prices will likely be higher as a result of decreased availability. Food will become less affordable, especially for the poor, causing social unrest.

Higher rates of obesity and undernutrition are probable because many people will have few options other than to eat less expensive, less nutritious, less perishable foods with empty calories.

In the face of climate change, it's increasingly urgent to realign our diets to focus on health and environmental sustainability and decrease food loss and waste while also fortifying nutrition. Individual and national purchasing and eating patterns will need to change. People will need to buy less food, and they will have to exercise greater food consciousness to reduce spoilage and waste.

These changes may seem difficult, but we should not fear or dread having to make these adjustments in our diets and habits. On the contrary, adapting in these ways would improve human health and well-being for everyone as well as planet Earth's prospects for generations to come.

THE INDUSTRIALIZATION OF AGRICULTURE

Many of the dramatic effects of food systems on climate can be traced to the industrialization of agriculture. Modern agriculture emerged on a global scale in the mid-twentieth century, when the Green Revolution spurred significant advances in the productivity of staple grain crops, especially in low-income countries. This agricultural transformation helped meet the needs of a growing population in those countries.

Renowned US agricultural scientist Norman Ernest Borlaug won both the Nobel Peace Prize and the World Food Prize for his work on the Green Revolution. He started the project in Mexico in 1944, and in tandem with the work of Indian geneticist M. S. Swaminathan, the revolution rolled through India and the rest of South Asia. With the benefit of hindsight, more contemporary food and agricultural scientists believe Borlaug's technological innovation (with Swaminathan) produced stronger, high-yielding varieties of staple crops, thereby saving millions of lives and avoiding widespread famine in India. Others wonder, however, whether dire consequences—including pollution, soil erosion, overdependence on fertilizers, and pressure on water tables—resulted from this ambitious revolution. Some argue that innovations also marginalized many kinds of foods and crop varieties while forcing communities to adapt to monocropping agricultural systems.

As with other industrial practices, industrial agriculture is based on scale and the segregation of tasks. Farms operate like efficient factories, deriving productivity gains from extreme specialization. Industrial agriculture relies on a limited number of crops and on synthetic fertilizers and pesticides. It seeks to deliver universal food security by producing massive volumes of uniform crop commodities for global markets. In turn, governments have supported the industrialization of agriculture through production subsidies, energy subsidies, and liberal trade regulations.

For the past half-century, agricultural development around the globe has focused on maximizing crop yields. During that time, global cereal production has more than doubled, outpacing the rate of population growth and averting widespread food shortages. In the United States, corn yields have increased more than eightfold since the 1930s, while soy and cotton yields have more than quadrupled.

To be successful, industrial agriculture requires large up-front investments in equipment, personnel training, marketing networks, and retail relationships. To make these investments, farmers generally must scale up their production practices and grow high volumes of uniform, commodity crops. Once farmers invest in industrialized practices, transitioning to a different system that may produce more nutritional diversity on farms is difficult.

Years ago while I was visiting Purdue University I visited some large monocropping farms in Indiana. One farmer who I met was a businessman who grew corn and soy and raised hogs. For him to introduce growing tomatoes for further diversification—what he called "boutique crops"—was incredibly risky. Making the change would have required him to understand immigration hiring laws (tomatoes must be handpicked) and buy special equipment to sort and sift the produce (tomatoes must be of a certain size and weight). It would have involved that he throw out half of the crop (customers don't buy "ugly" tomatoes). This farmer's anticipated return just wasn't worth his up-front investment.

CHANGING AGRICULTURE LANDSCAPES, CHANGING DIETS

More intense focus on production has spurred economic growth while also increasing food security in many regions of the world. This shift, however, has also compromised both nutritional (for humans) and environmental (for planet Earth) health. Many of these negative effects stem from the model's uniformity. Industrial agriculture typically causes a dramatic loss of genetic diversity in crops and animal populations. National food supplies became larger (people are eating more food than they did 50 years ago), with globally important staple foods (wheat, rice, maize, and sugar) making up the majority of the supply. Other crops have emerged as widespread staples, particularly soy, palm, sunflower, and rapeseed used for oils.

As these crops become more prevalent in food supplies around the world, alternative traditional staples such as sorghum, millet, rye, cassava, sweet potatoes, and yams have been marginalized. They haven't disappeared—at least not yet—but they've become less prevalent as foods eaten every day in many places. Traditional diets based primarily on singular staples (for instance, rice in Southeast Asia) have changed over time to include more common staples such as wheat and potatoes. The same is true for maize-based diets in Latin America, sorghum- and millet-based diets in sub-Saharan Africa, and so on around the globe.

Food supplies worldwide are much more similar and among crop species there are fewer varieties available today than there were a century ago.[3] Colin Khoury, a crop diversity specialist at the International Center for Tropical Agriculture in Colombia, explored this topic. He argued: "If we are what we eat, then it seems that over the last half-century people around the world have become much more the same type of human being—globalized people eating globalized foods."[4]

Another important aspect of agriculture is the subsidy policies that support economic growth. In the United States, the impact of agriculture subsidies on nutrition is highly debated and lacks clear evidence to support one perspective or another. Older studies indicate that overproduction of subsidized corn and soy is one of the primary causes of increased consumption, which contributes to obesity in a population. Others argue, however, that current subsidy policies in the United States have a minimal, if any, impact on dietary patterns and obesity. Regardless, most agriculture subsidy programs contribute funding to commodities that often don't align with dietary guidelines. And even further, dietary guidelines rarely consider environmental sustainability.

As part of agriculture's industrialization, specific supply chains, such as those that produce farm animal feed or processed food ingredients, have come to rely on export markets. Such reliance exposes these supply chains to price volatility, trade embargoes and tariffs, and sourcing issues due to environmental degradation. Nonetheless, government policies continue to support these supply chains over other more nutritious supply chains because of their massive economic and caloric potential. The result is various food systems in which only a limited number of crops and producers are in play.

Industrial agriculture is resource intensive and significantly contributes to the world's greenhouse gas emissions. It exacerbates land clearing and relies on high inputs of energy, fertilizer, pesticides, and water. It can under certain contexts, damage biodiversity and natural ecosystem cycles, even though uniform crop cultivation can increase yields. A more diversified form of agriculture could provide more resilience and protection against disaster, but the time and costs needed to rebuild soil health and fertility dissuade farmers from taking this route. At the same time, future agriculture expansion will be curbed by urban expansion. It's expected that urbanization will result in an approximate 2 percent loss of global croplands by 2030. About 80 percent of that cropland loss will take place in Asia and Africa.[5]

Ultimately, evidence suggests that despite increased yields, the abundance of the world's major crops such as maize, rice, wheat, and soy have begun to stagnate.[6] It is possible that crop productivity can be increased only up to a certain point, after which new techniques are necessary for a boost. Combined, these factors support moving away from industrialization, or at least rethinking our intensification efforts to better support human and environmental health.

A WARMING PLANET

Even as crop yields have begun to stagnate, the Earth has gotten warmer, and temperatures will continue to rise. The human species is altering the climate in ways that are leading to dire consequences.

On our current trajectory, the world will warm 1.5 degrees Celsius more from preindustrial levels sometime between 2030 and 2052, and 3.2 degrees Celsius of warming by the end of the century.[7] In a worst-case scenario of accelerating emissions, areas currently home to a third of the world's population will be as hot as the hottest parts of the Sahara within 50 years. Even with the most optimistic outlook, 1.2 billion people will reside outside the comfortable "climate niche" in which humans have thrived for at least 6,000 years.[8] This, along with rising waters that drown coastal areas, will cause people to migrate to more livable places on the planet, increasing population pressure in some parts of the world.

Even with drastically reduced greenhouse gas emissions, if our food systems remain on their current course they will likely lead to 1.5 degrees Celsius of warming by the end of the century. The projected warming of the planet will result in more hot days and *hotter* hot days around the globe, with regions along the equator becoming unsafe for human health. Sea levels will rise. Biodiversity will be decimated. Coral reefs will all but disappear, and marine fisheries will see yields drop. Climate-related effects will pose threats to all people and all nations.

Even more troubling is the possibility that the planet could reach global tipping points that cause entire Earth systems to collapse. Such tipping points might include the loss of the permafrost layer, the loss of the Amazon rain forest, the melting of the West Antarctic or Greenland ice sheets, or cessation of major ocean currents. The result would be a complete loss of human sovereignty over Earth systems. My friend Johan Rockström (or Johan Rockstar as I call him), director of the Potsdam Institute for Climate Impact Research in Germany and professor in Earth System Science at the University of Potsdam, argues that "deforestation and climate change are destabilizing the Amazon—the world's largest rain forest, which is home to 1 in 10 known species. Estimates of where an Amazon tipping point could lie range from 40 percent deforestation to just 20 percent forest-cover loss."[9] Sadly, 17 percent has already been lost since 1970. And we won't get it back.

Regions will not suffer equally from climate change. In some areas, such as the Andes and East African highlands, growing seasons may expand. The production of cassava is projected to increase with climate change because cassava trees (which produce edible roots and leaves, a staple crop for many Africans) thrive in warmer temperatures and respond positively to carbon dioxide increases. However, in most of the Global South, particularly in equatorial regions, climate change is expected to decrease various crop yields and alter where many foods can be produced.

For regions already at risk of food insecurity, climate change could further decrease food availability and increase food prices. Cyclone Idai caused widespread food insecurity when it hit Mozambique in 2019. This cyclone is considered one of the worst climate-related disasters in the Southern Hemisphere. After the cyclone hit, I was watching a video on the *New York Times* website in which the very first thing a woman from Mozambique said was: "We don't have any food that we're going to be able to eat tonight." More natural disasters loom, and they will have graver consequences for humans, for their food security and diets, and for their homes, more than ever before.

Most of the world's acute hunger and undernutrition occurs not during conflicts and natural disasters but during annual "hunger seasons"—the times of year when the previous harvest's stocks have dwindled, food prices are high, jobs are scarce, and rainfall is unpredictable. The frequency and intensity of seasonal hunger are expected to increase with climate change and to be especially severe in Africa south of the Sahara, causing shocks to food systems. Such effects will be most dire in areas where agriculture is rainfed and rains are highly seasonal.

Food shocks will continue to happen. COVID-19 was initially thought of as an acute health-related shock but has evolved over time into a long-term shock to multiple systems, including health care, the economy, and food supply chains. Climate change is a long-term shock to food systems. Some argue that extreme weather events such as heat waves, droughts, flooding, and cold spells can lead to devastating failures of major crops, including wheat, maize, soy, and rice. The risk of extreme weather events co-occurring at multiple global cropping locations is increasing because of climate change.[10] "Multiple breadbasket failures," as some call this phenomena, are likely to occur in the next two decades, compromising the ability of billions of people to access food.

Agricultural production involves a feedback loop with the environment. It contributes to ever-increasing climate change, and climate change drives intensified production to meet the global food demand. Fossil fuels are one component of this feedback loop. They're used to produce fertilizers, pesticides, and synthetic agrochemicals, which significantly increase crop productivity. At the same time, they have negative environmental consequences—groundwater contamination, soil acidification, soil biodiversity loss, and buildup of chemicals in the waterways and on the land—which can be toxic to humans and animals. These adverse outcomes could lead to decreased crop yields in the long run, which in turn require even more chemicals to increase productivity. This vicious cycle will be incredibly challenging to break.

Maintaining, much less expanding, agricultural production will become increasingly difficult in the face of hotter temperatures, a more limited water supply, and the acidification of the soils and oceans. Heat-stressed plants are more susceptible to disease, which could lead to decreased yields and increased use of agrochemicals for pest control. Some pest populations are expected to flourish in warmer temperatures and migrate to new, higher latitudes.

As a result of our changing climate, less food will be available and the quality of food in general will diminish. Elevated levels of carbon dioxide in the atmosphere can increase photosynthesis rates and growth. This increased growth, however, can also reduce some crops' nutritional value, especially for wheat, rice, potatoes, soy, and peas. Samuel Myers, the director of the Planetary Health Alliance at Harvard University in Cambridge, Massachusetts, and an expert on the human health impacts of global environmental change, has shown that greater productivity gains may offset the yield-decreasing impact of climate change, but the harvested crops typically contain less protein, iron, and zinc, essential nutrients for human health.

According to one estimate, unless dietary patterns change, diets higher in meats, refined sugars, fats, and oils, could contribute to an 80 percent increase in agricultural greenhouse gas emissions and global land clearing by 2050.[11] Environmental degradation will lead to increasingly constrained and disrupted agricultural practices and access to natural resources and ecosystem services, with the planet no longer able to sustain its current methods and intensity of food production.[12] The resulting constraints to diets will pose threats to human nutrition and health more alarming than those that exist today.

What Shocks Global Food Systems?

Food systems are composed of production, processes, and infrastructures required to feed a population. These include everything from plant and livestock breeding to growing and raising crops and herds, processing them, packaging the foods made from them, shipping them, consuming them, and handling the organic and packaging waste that remains. Within those steps, hundreds of elements are in play, including policy agreements, workforce conditions, irrigation techniques, pesticide inventions, animal welfare innovations, preservation tools, transport and distribution technologies, marketing campaigns, nutritional benchmarks, and recycling efforts.

While there are routine failures throughout these steps, catastrophic natural and man-made shocks to the systems also occur, including:

- wars and other armed conflicts that impact labor forces and transport of goods;
- extreme weather events, such as droughts and hurricanes;
- shortsighted farming practices, such as the overplowing and removal of drought-resistant grasses that contributed to the 1930s Dust Bowl in the American and Canadian prairies, and the deforestation in the Amazon to clear land for cattle grazing and soy farms that fed the fires of 2019–20;

- crop failures—infamously, the potato fungus blight that forced millions of Irish to flee their homeland in the nineteenth century or face starvation from the famine that killed one million of their countrymen between 1846 and 1851;
- insect infestations, such as the trillions of locusts that decimated pastures and crops in Africa and India during the summer of 2020, and large-scale destruction of grain stores from a variety of bugs throughout history; and
- pandemics, including COVID-19 that kept workers from the fields, kept ships in harbors, sickened workers who were forced to remain in meat processing centers, and caused food shortages around the world.

These shocks wreak havoc on both the food supply chain and the environment and can displace entire populations for months, years, or decades, as well as kill many thousands who starve or succumb to hunger-related illnesses. In the case of the Dust Bowl, Midwesterners fled to California in droves, and Ireland has never recovered its prefamine population levels following its mass emigration more than a century ago.

DECLINING AGRICULTURAL DIVERSITY

One consequence of industrialization has been a steady decline in the diversity of plants and animals used in agriculture.[13] When I worked at Bioversity International in Rome, I led its nutrition portfolio, exploring how we can sustainably use and conserve biodiversity on the planet for diets and nutrition. I collaborated with young scientists such as Roseline Remans and Fabrice DeClerck who have gone on to push the scientific frontier to address how biodiversity and ecosystems are critical for human health and well-being—as well as the prospects for planet Earth—for generations to come.

During this time at Bioversity International, we collaborated with Crop Trust, which manages the Svalbard Global Seed Vault in Norway, also known as the "Doomsday vault." This fail-safe seed bank is a safety deposit box for the world's genomic diversity of crops. It's meant to store and preserve the many varieties of seeds of different food species in case any are wiped out by natural or man-made disasters. The vault currently protects around 930,000 seed samples, representing 5,000 plant species; its storage capacity is 4.5 million seed samples.

Throughout human history, people have used roughly 7,000 plant species as food sources along with a wide variety of animals and other species, including fungi, algae, yeasts, and bacteria. Over the past century, however—and primarily by conscious choice—humans have driven massive declines in the diversity of agricultural systems. Of the more than 50,000 edible plants on Earth, only 15 crops are used to meet 90 percent of the global population's caloric demands. Even more noteworthy, only three staple crops—rice, maize, and wheat—account for two-thirds of global food energy intake. A century ago, commercial seed houses offered hundreds of varieties of crops that provided nutritional diversity, risk adversity, and climate adaptability.

Today, the Food and Agricultural Organization of the United Nations argues that the world's agricultural landscape is dominated by only 12 species of grain crops, 23 species of vegetable crops, 35 species of fruit and nut crops, and 5 animal species (this doesn't include fish which will be discussed later). Globally, 75 percent of the land used for agriculture is devoted to growing these 12 crops. In India, more than 80,000 varieties of rice were once cultivated, but that number has fallen to just several hundred.[14] Similarly, the United States has largely shifted to monocultures of corn and soy, with the great majority of farms producing the same varieties of the same crops. This creates incredible risk not only from a nutritional perspective but also from a climate perspective. As with an investment portfolio, it pays to diversify, a hard-earned lesson of the Irish Potato Famine.

Although consumers generally have more choices among plant-based foods than they once did, the total diversity of crops that make significant contributions to our diets has dwin-

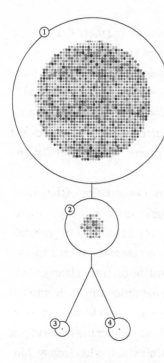

① **391,000**
GLOBALLY IDENTIFIED
PLANT SPECIES

② **5,538**
NUMBER OF CROPS USED
FOR FOOD BY HUMANS
THROUGHOUT HISTORY

③ **3**
RICE, MAIZE, AND WHEAT
CURRENTLY PROVIDE >50%
OF THE WORLD'S CALORIES
FROM PLANTS

④ **12**
12 CROPS THAT TOGETHER
WITH 5 ANIMAL SPECIES*
PROVIDE 75% OF THE
WORLD'S FOOD TODAY

*(IN ORDER OF GLOBAL
CONSUMPTION, COWS, CHICKENS,
PIGS, GOATS, AND SHEEP)

BIODIVERSITY DECLINE: Of all the plant species identified around the world, only about 1 percent are cultivated for consumption. Twelve crops and five animal species make up 75 percent of our food supply. Those 12 crops include sugar, maize, rice, wheat, potatoes, soybeans, cassava, tomatoes, banana, onions, apples, and grapes (wine). The five animal species are meat and by-products from cattle (such as milk), chicken (and their eggs), pigs, goats (and milk), and sheep (and milk). Animal farming also contributes to a loss of biodiversity among livestock; between 2001 and 2007, 62 livestock breeds became extinct for myriad reasons including their unsuitability to industrial production.

© Bioversity International (2020).

dled. Many factors have contributed to the decline, including replacing human labor with machinery and investing in breeding and distributing high-yielding major crops as a development strategy. In addition, agriculture subsidies dedicated to a narrow range of crop commodities have also contributed to reduced diversity in the global food supply. This trend toward homogeneity in the global food supply also heightens interdependence among countries with respect to the availability and access to vital foods as imports.[15]

As farm sizes increase, the diversity of crops produced and their nutrient content typically diminish. Smaller farms with more agrobiodiversity often introduce a broader array of nutrients (particularly micronutrients) in the global food supply than do large farms.[16] Mario Herrero who led that research said: "Small and medium holder farmers are providing a monumental ecosystem service. They're the stewards of the nutrients and biodiversity for the world." However, despite the arguments for diversifying, smallholder farms (generally defined as less than 2 hectares, about 5 acres, but sometimes defined as up to 10 hectares) are the most vulnerable to the effects of climate change. They also are the most disenfranchised from the global financial system. Many smallholder farmers, especially women, struggle to emerge above subsistence levels. They often lack access to credit, technical support, and markets while enduring the volatile price swings of global commodity markets. Smallholder farmers rely on additional sources of income to survive, sometimes outside of farming, a backup strategy that may become increasingly important in the face of climate change.

I saw this firsthand in Malawi and Mali, countries on opposite sides of the massive continent of Africa. Subsistence farmers try to eke out a living by growing crops to send to market, with enough food left over to feed their families. Unfortunately, these small farms often fail because of droughts, unpredictable rains, lack of mechanization or technology to support their small business, and no infrastructure (sometimes not even roads) to get their crops to distant markets. Even with the odds these farmers face, their land still contributes 30 percent of all food commodities in their regions.

Globalization has intensified downward price pressures and costly regulatory burdens for farmers of all sizes. As a result, half of the hungry people in the world are fed from small-scale farming communities.[17] It's ironic that those who feed us are often the most food insecure. Nearly one billion people who derive their livelihoods primarily from agriculture are the populations that will bear the brunt of large-scale environmental change in the near future. Climate change may even cause smallholders to abandon their own farms in an effort to find secure food sources and livelihoods.

The use of pesticides in agriculture compounds the downward spiral of dwindling crop diversity. Pesticides have reduced the numbers and diversity of species of pollinators such as bees, bats, and butterflies, which play vital roles in the production of fruits, vegetables, nuts, and seeds.[18] In general, climate change poses a huge threat to the survival of all insects, which has vital implications for human environments. Insects not only pollinate crops and flowers but also provide food for higher-level organisms, break down detritus, maintain a balance in ecosystems by consuming leafy plants, and help recycle nutrients in the soil. Bee populations have been declining because of colony collapse disorder, a poorly understood phenomena that may be the result of habitat loss, use of pesticides and fertilizers in agriculture, and/or the Varroa mites that feed on bees.

Beyond its effects on diversity, intensive agricultural production contributes to diminished soil quality, which is crucial for the micronutrient content of crops. Physical, biological, and chemical degradation of soil caused by intensive agriculture exacerbates the negative effects of weather and climate on soil stability and quality. Erosion and the degradation of soil quality have already resulted in the depletion and subsequent abandonment of roughly one-third of the world's arable land.[19] The south-central United States experienced the dire consequences of poor soil conservation practices in the 1930s with the Dust Bowl, which contributed to the Great Depression and forced 2.5 million people to migrate across the country. Increased use of techniques to reduce soil erosion, such as no-till methods and cover crops, will be necessary to meet the global food demand and avoid repeating such catastrophes.

Industrialized agriculture aims to fulfill nutritional needs by combining highly specialized and productive farming with well-functioning trading systems that allow consumers to buy a variety of crops. In execution, however, the diversity of foods delivered by international trade has mainly benefited wealthy consumers in high-income countries. Meanwhile, substandard infrastructure and broken or inadequate value chains have forced poor people in low-income countries to rely on staple crops that are insufficient to meet their nutritional needs.

WATER PRESSURES

Climate change will dramatically affect the use of water by agriculture. Water availability will become more unpredictable even as flooding, droughts, and sea levels increase. Rural areas in particular can expect major repercussions in water availability, food security, infrastructure, and agricultural fallout.

Groundwater currently supplies roughly half of the fresh water used domestically around the world, but in many areas it's being extracted faster than it can be replenished by rainfall.[20] At the same time, climate change is exacerbating the variability of precipitation, resulting in atypical rainfall patterns that involve droughts, floods, and storms. In Australia, recent droughts have been especially severe. Future droughts are expected to be more regular, longer in duration, and broader in the areas impacted.

While Australia faces worsening droughts, areas of Bangladesh suffer from increased flooding and sea-level rise. Intensifying storms caused by climate change are not only destroying homes and livelihoods but also contribute to higher water and soil salinity while leaving millions of people with little to eat or drink. According to the Intergovernmental Panel on Climate Change, by 2050 rising sea levels caused by climate change will submerge roughly 17 percent of Bangladesh's land, displace approximately 20 million people, and result in a 30 percent reduction of food production in the South Asian country.

Changes in the frequency, duration, and intensity of rainfall can produce a range of results. An increase in rainfall could support food production, especially in sub-Saharan African countries where advanced irrigation technologies are uncommon. Increased rainfall, however, could also damage crop output. Large storms and flooding have been shown to decimate crops in India and in the last few years, in Iowa and Nebraska, the breadbasket of the United States.

Innovative measures will become increasingly necessary as more regions experience water variability from extreme rainfalls or droughts. During my time in Kenya, I witnessed many changes to food production systems designed to combat the effects of climate change. I worked in the village of Dertu, near the border with Somalia. Rainfall has been highly erratic throughout the region in recent years, and these effects have been experienced deeply in Dertu, a largely pastoralist (herder) community focused on raising camels, cows, and "shoats" (sheep and goats). I was amazed at how far pastoralists walked with their animals from one water borehole to the next, grazing in between—as much as 50 kilometers or more in a day. Their livelihoods are precarious because of climate change, and many pastoralists face a difficult future because of water and food insecurity, landownership disputes, and droughts.

To improve the production of milk despite recurrent harsh droughts, Dertu and neighboring villages vaccinated their shoats and camels to protect the animals during stressful climatic conditions, including drought and unexpected floods that change the parasitic milieu. Other initiatives have also been implemented to improve food security and the livelihoods of pastoralist communities amid climate change. Communication technologies, including mobile phones and solar chargers now connect these remote villagers to the outside world and provide information about weather, security, livestock market prices, status of boreholes, and water surface availability. This information makes it possible for these communities to respond more proactively to droughts.

The Dertu Renewable Energy Project has brought biogas (from livestock manure) to the community, while solar power is being used to create new business enterprises. One such enterprise is producing high quality camel milk that can be refrigerated and transported to distant supermarkets. Camels are highly valued in the Horn of Africa and camel milk has been a part of the Kenyan Somali food systems culture for millennia. In Nairobi, camel milk has become a popular alternative to cow's milk. Vitamin C levels in camel milk are three times higher than in cow's milk, which makes it particularly important in harsh regions, where diets often lack fruits and vegetables. Camel milk is also rich in iron, unsaturated fatty acids, B vitamins, and the protein lactoferrin, which has antibacterial properties.

FISHERIES AND CLIMATE CHANGE

Fisheries are vital to global food systems and diets. Fish provide protein, omega-3 fatty acids, and micronutrients to many populations. For 4.3 billion people, fisheries account for at least 15 percent of total protein consumption.[21] Moreover, coral reef ecosystems provide food and other resources to 500 million people worldwide.

Nearly half of the global population lives within 100 kilometers of the coast, yet half of these coastal dwellers have moderate to severe micronutrient deficiency risks that could be significantly reduced if fish were more readily accessible. Many nutrients are available from the fish already caught, but these fish catches are not reaching many local populations that often need these nutrients most. The amount of fish currently caught off the West African coast is sufficient to meet the nutritional needs of the people living within 100 kilometers of the sea. These people, however, are not benefiting from this supply and continue to suffer from zinc, iron, and vitamin A deficiencies because of access barriers due to international trade priorities (these fish garner significant income overseas), cultural food preferences, food waste, and the reduction of whole fish to fish oil for animal feed.[22]

Unfortunately, aquatic food sources also face increasing threats from overfishing and climate change. Overfishing has decimated ocean, river, and lake populations, and reef degradation has led to a decline in the abundance of fish and invertebrate species. Fish farms have contributed to pollution and the spread of disease. Climate change has altered ocean temperatures, salinity, oxygen, and acidification as well as freshwater temperatures and water levels. Absorption of carbon dioxide by the oceans results in more acidic water, which threatens fish survival, drives shellfish degradation, and causes coral reef bleaching. Plus, ocean acidification caused by cumulative greenhouse gas emissions over several decades is threatening single-cell phytoplankton that form the base of marine food chains and account for more than half of the photosynthesis and oxygen production on Earth. As oceans warm, fish species are migrating away from the tropics toward cooler water, and more will be forced to do so as water conditions become increasingly uninhabitable. Already, ocean warming has led to a 40 percent decrease in fish in the tropics and a 30 to 70 percent increase at the poles.[23]

Captures from ocean fisheries peaked in the 1990s. Currently, 57 percent of fish stocks are fully fished out, and 30 percent are overexploited and overfished. Overall, approximately 90 percent of fish from monitored fisheries are harvested at or above maximum sustainable yields. Meanwhile, more than 60 percent of the world's rivers have been dammed to control water resources and produce energy. With the wild-caught fish global supply past its peak, the world's fish consumption will need to be supplied by farmed fish, also known as aquaculture. The aquaculture sector will need to double its ability to feed the world by 2050.

Despite these daunting figures, the potential exists for steady improvements in the aquaculture sector that could instead contribute to its future growth. These improvements include better breeding and hatchery technologies that don't impact surrounding ecosystems, more sustainable feeds (some involving seaweed), and better disease management. Innovations with recirculating aquaculture systems could lead to efficient water and nutrient use, possibly offsetting some of the sustainability challenges that the aquaculture sector currently faces. Consumers too could shift toward seafoods that are lower on the food chain, including bivalve mollusks (for example, clams, oysters, and mussels), catfish, and carp.

BRINGING TECHNOLOGIES TO CROPS

Technological innovations have transformed food systems over the past century. Plant and animal breeding, mechanization, agricultural chemicals, irrigation, and many other technologies have enabled farmers to produce more food using less land than ever before. However, many of these same technologies have contributed to worsening health and environmental outcomes or have not reached all those that could benefit from the technologies.

Some of the most contentious issues involve genetically modified (GM) food. GM crops, which were first commercialized in the United States in the 1990s, originally were designed to have two advantageous traits: (1) resistance to insects or diseases and (2) tolerance to herbicides. More recently, GM crops have been approved for other traits, such as drought tolerance and nonbrowning, and researchers are working on such possibilities as increased photosynthetic efficiency, removal of allergens, tolerance to temperature extremes, disease and pest resistance, and improved taste, aroma, and nutrition.

GM crops have had a polarizing effect since their development. In 2016, a committee of the National Academies of Sciences, Engineering, and Medicine published a review of roughly 1,000 studies about the safety of GM crops and concluded that they are safe for human consumption.[24] However, other concerns have arisen. One is that the technology will compete with other strategies to improve nutrition and reduce environmental harm. Another is that GM crops do not necessarily increase crop yields and can instead increase the use of chemical pesticides and threaten pollinators. In addition, there are concerns that

GM crops have the potential to displace traditional crop varieties that have higher nutrient value and are locally adapted to the environment.

The productivity gains made possible by GM crops have provided some farmers with more income. In 2014, a review of 147 studies reported that on average GM crops increased agricultural productivity by 22 percent and increased farmers' profits by 68 percent.[25] In 2016, a study found that for each dollar invested in GM crop seeds, farmers in low-income countries gained an average of $5.[26] Today, more than 18 million farmers a year plant GM crops.

GM technology is also changing agriculture's carbon footprint, depending on the crop and how and where it is grown. GM crops have the potential to allow farmers to produce higher yields without needing to use additional land, alleviating some of the pressure to convert more finite land for agricultural production. Moreover, research has found that micronutrient-rich seeds are associated with soil enrichment, thereby improving agriculture's contributions to environmental health and with less use of insecticides and herbicides (such as glyphosate). A recent study found that, in the United States, GM maize used less herbicide, but GM soy used more herbicide.[27] What researchers did find is that glyphosate-tolerant variety adopters for both crops saw an emergence of glyphosate weed resistance over time. The problem is that farmers will use more of the herbicide to kill weeds, which could be detrimental to the environment. There are also concerns about glyphosate exposure on human health, and studies are ongoing to examine its toxic effects and potential risk for cancer and other deleterious effects.[28]

Biofortification, which may or may not rely on genetic technologies, is another tool that has great potential to reduce global malnutrition. Biofortification is the process by which the nutritional quality of food crops is improved by adding nutrients or other health-promoting properties through agronomic practices, conventional plant breeding, or modern biotechnology. Howdy Bouis, who started HarvestPlus, an organization focusing on developing the technologies of biofortification, told me that "looking to the future of biofortification, I very much hope that genetic engineering can be as widely and easily used as conventional breeding techniques. Multiple nutrients and agronomic improvements can be added simultaneously to further improve popular crop varieties, and higher nutrient densities are possible. Proofs of concept of these advantages are already available. These techniques are proven safe and have been endorsed by academies of science all over the world. Investments in agricultural research have among the highest documented benefit-cost ratios."[29]

Current biofortification efforts focus on staple crops such as corn, rice, potatoes, and wheat since these crops often supply the majority of people's calories in low-income countries. Biofortification is designed to improve the nutritional content of the staple foods that people already eat, providing a comparatively cost-effective and sustainable means of addressing micronutrient deficiencies among poor populations. Studies have shown that biofortified crops are at least as productive, in terms of both yield and economic value, as their traditional counterparts.

Opponents of biofortification worry that these nutrient-enhanced crops will dominate poor people's diets with less effort to expand dietary diversity beyond staple grains. Biofortified crops could also displace some traditional crop varieties that are nutrient rich and adapted to the local environment.[30] In addition, farmers worry that biofortified crops might distort local market prices for agricultural commodities and thereby threaten their livelihoods. As a result of these concerns, some biofortified crops in certain settings have faced barriers.

POSTHARVEST STORAGE AND PROCESSING

Food is stored and processed to make it more stable, safe, and, in some cases, nutritious. Storage is a huge issue in many low-income countries because of outdated, shabby infrastructure; sometimes crop harvests are lost due to pest infestations, mold, and other issues. In some East African countries, aflatoxin, a toxin found in soils, can destroy 50 to 60 percent of harvested maize and peanuts. While I was in Mbola, Tanzania, in 2010, I visited a storage facility where bags of maize were being stored for a school meal program. A weevil infestation turned the maize to powder. The loss was tragic for the villagers. Silos and Cold-Hubs are now being tried in many places to protect crops during storage.

Common food processing practices include milling, cooling or freezing, smoking, heating, canning, fermenting, and extrusion cooking, all of which have environmental consequences. Cookers, boilers, and furnaces emit carbon dioxide. Wastewater emits methane and nitrous oxide. The most intensive processing method is the wet milling of maize, but processing sugar and oils also requires large amounts of energy. In addition, resources are required to produce packaging materials and for the packaging process, although these energy contributions are minimal compared with the rest of processing. Post-consumption waste from packaging is also an environmental concern.

Processing practices can alter the nutrient content and absorption of that nutrient within food. Some techniques enhance the nutritional value of foods and extend their shelf life. Fermentation processes can preserve food for times of scarcity while also imparting desirable flavors to and reducing toxicity in foods. Lactic fermentation of vegetables, such as sauerkraut or kimchi, adds nutritional and microbiological diversity to the diet, which can improve microbiome health because of prebiotic properties. Other techniques, however, such as reformulation, may increase or decrease the nutritional value of food by removing fiber and other key nutrients, or remove unhealthy ingredients like sodium and added sugars. Conversely, some techniques can compromise health by *adding* unnecessarily high levels of ingredients that typically should be limited, such as sodium, sugar, and unhealthy fats, such as trans fats.

After processing, many foods go into cold storage prior to distribution. Refrigeration is one of the most important breakthroughs in food supply chains because it reduces food spoilage and waste and enhances food safety. Refrigeration has a clear downside, however: Second to production, it is the most energy-intensive step in the food value chain. In addition, it uses refrigerants that can damage the ozone layer.

Fortunately, innovative techniques to improve storage and processing are being developed. One promising method to strengthen the sustainability of food production systems is called upcycling, which makes use of the by-products of food processing. An example is the use of sweet whey, a by-product of cheese making. Originally considered a waste, whey is now used as an ingredient in many "health food" products, and the cream skimmed from whey is used to make whey butter, an ingredient in butter-flavored foods. Upcycling can be lucrative for food manufacturers, can increase resource efficiency, and can contribute to sustainability management plans.

DISTRIBUTION, MARKETING, AND RETAIL

Once food has been processed, it's sold through formal or informal markets that may be near or far from the communities and households they serve. The international food trade has experienced rapid growth over recent decades due to the development of refrigerated cargo and container ships. Today, seasonal foods are increasingly available to people all over the world willing to pay the price for them—and accept their carbon footprints.

This movement of food commodities across international borders is changing food consumption patterns worldwide. Quinoa, an ancient nutrient-rich grain found traditionally in the Andes Mountains, is a new niche commodity in international trade. With growing demand from markets around the globe, quinoa production more than doubled in Peru between 2011 and 2014, mainly through an increase in yield per hectare. Peruvian farmers have benefited economically from this increased demand, but intensified production has also resulted in more environmental harm and higher domestic prices.[31]

Climate change will increasingly affect producers' ability to move food from production to markets. The fallout includes making access to diverse high-quality diets more difficult and contributing 6 percent more to total greenhouse gas emissions coming from the food system. Food is transported via water, rail, road, or air, all of which require fossil fuels that contribute to greenhouse gas emissions. Air transportation contributes the highest level of emissions, while railroads contribute the least, depending on distance traveled. Fortunately, most food travels by sea freight, which is cheaper than other options and emits 50 times less greenhouse gases than transporting the same amount by air.

At the same time, as urbanization in low-income countries increases, the distance food must travel from the site of production to the site of consumption is increasing. Urban demand will increasingly dictate what foods are grown by rural producers and how these foods are traded, processed, distributed, and marketed.

By 2050, an estimated 66 percent of the world's population will be living in urban centers. Although Africa and Asia currently remain predominantly rural, urbanization is occurring in these nations faster than throughout the rest of the world. Urbanization will put added stress on food systems through heightened consumption and demand for a larger array of foods. In addition, smallholder farmers are abandoning their agricultural lifestyles in favor of city living.

At the tail end of global food systems, massive food waste adds to resource depletion and climate change. Of the approximately 4 billion tons of food produced annually, roughly one-third is wasted.[32] In low-income countries, food loss occurs primarily during the production-to-processing stages of the food supply chain. In higher-income countries, most food waste occurs at the retail and consumer levels. Relatively more staple crops, fruits, and vegetables are wasted than animal products. As former Secretary of State Madeleine Albright wrote: "In a world where one-third of all edible food never makes it to the mouths of the hungry, we all have an individual moral responsibility to do our part."[33]

A growing world population will require more food. At the same time, climate change will make it increasingly difficult for farmers to feed our population. Our current food systems focus on cheap, abundant food that can be made at maximum speed, with good profit margins for those industries that dominate the marketplace, leaving many smallholders behind. Weathering climate change will require much more sustainable approaches. Without a significant restructuring of food production practices, the effects of climate change will reduce the food security and nutrition of the world's population. With urban encroachment on less populated locales and 60 as the average age of the world's farmers (among those counted), the question becomes, *Who will feed us?* And who would *want* to be a farmer with the daunting prospects of climate change barreling down on us?

We must determine how to meet the world's caloric and nutritional needs while minimizing further harm to the planet, and while ensuring farmers have the support they need to adapt to a changing climate. The past century has been one of agricultural intensification and uniformity. To improve human and environmental health, the next century requires a new approach.

CHAPTER 3

Do We Have the Right
to Eat Wrongly?

MANY PEOPLE DON'T THINK OF EATING as an ethical act, but the food choices we make and the systems that supply our meals raise far-reaching and difficult ethical dilemmas. We need to collectively grapple with and resolve these issues if we want to ensure that everyone has equitable access to healthy and sustainable diets.

We can start by asking ourselves some difficult questions. Do we have a right to eat resource-intensive foods such as beef that also represent ethical quandaries about what animals, if any, we should farm to eat? Should governments have the power to tell people what to eat to improve their health? Can women be properly nourished if they eat last in their households or don't have equal control over household income? How can we address historic and systemic inequities and discrimination to ensure healthy diets are available regardless of wealth, residency, skin color, tribe, or caste? On the global balance sheet, what do wealthy nations "owe" impoverished nations that struggle to grow food because of climate change caused largely by the actions of industrialized nations and multinational corporations?

Worldwide, many people still struggle to procure adequate food to meet their basic needs—a daily concern that becomes a crisis when there are droughts, pandemics, wars, or other natural or human shocks to food and other linked systems. For others, the types and amounts of food that they have access to are unhealthy and increase the likelihood they'll experience a disability and/or early death. Macro-level factors tied to agriculture, natural resources, infrastructures, and economics influence individual access to acceptable and affordable foods. Our entrenched food systems and some of the actors who steer them in different directions factor into why these inequities persist and even intensify. At the end of the day, food systems often marginalize populations by denying opportunities and limiting options. This leaves them disproportionately vulnerable to insufficient diets, malnutrition, and health burdens. Marginalized populations include the urban poor, the rural, the geographically isolated, women and girls, the discriminated, the disabled, those who live in conflict zones, those disproportionately affected by climate change, and now, the COVID-19 pandemic. The question is, why haven't these populations been more valued in society?

Food systems that prioritize equity emphasize well-being and social justice. These twin pillars include fair trade laws, fair income and safe working conditions for people in the food industry, improved food safety and protection for consumers, healthy and sustainable human diets in sync with environmental sustainability and stewardship, and accessibility of affordable food for citizens.

Almost everyone would prefer food systems that are sustainable, equitable, and healthy (for the planet as well as humans). But that's a big ask according to Tara Garnett, the Food Climate Research Network founder of the Oxford Martin Programme on the Future of Food and an expert on food systems and issues linking to sustainability and equity. Garnett sums it up: "Not everyone has the same vision on what the solution—the good life—might look like. The ethical perspectives people bring to the food-sustainability problem influence both their use of the evidence and the solutions they propose—and these often lead to stakeholders arguing at cross-purposes, the result being conflict, or inaction."[1]

It's impossible to highlight all the major equity issues associated with food systems and diets in a single chapter. Here, the focus is on inequities in accessing affordable healthy diets, the ways in which food environments are built and designed, the unsustainability of current animal production and consumption, and the marginalization of one of the most important key stakeholders in food systems: women.

INEQUITIES IN THE COST OF DIETS

Issues of undernutrition and obesity around the world have become much better understood in recent years, but the forces that drive these conditions have received much less attention. Food is not a cause of inequities. Rather, food inequities are a symptom of larger systemic issues, such as extreme poverty, exclusion, disability, exploitation, and social injustice.

Income is a critical factor in reducing malnutrition and improving health outcomes. Despite the economic contributions that food systems make to livelihoods around the world, healthy and diverse diets remain cost-prohibitive for many people.[2] Unhealthy diets slow economic growth and perpetuate poverty in three major ways: (1) They cause direct losses in productivity from poor physical health. (2) They produce indirect losses from decreased cognitive function and deficits in finishing school and fulfilling education potential, all of which impact the strength of workforces. (3) They create losses due to increased health care costs. These economic losses contribute to entrenched, difficult-to-break cycles of poverty.

In high-income countries, dietary inequity is closely linked to wealth: high-quality diets are associated with higher socioeconomic status, while unhealthy diets packed with calories but low in nutrients are more prevalent among lower-income groups, communities, and neighborhoods. One reason people eat low-quality food is that healthier diets tend to be more expensive because they're perishable and require extra care, including cold storage and distribution to get from point A to point B. In particular, animal foods and dairy products, vegetables, and fruits are prohibitively expensive for many people. In deeply rural sub-Saharan Africa, fruits and vegetables are seasonal, and when animal-source foods do reach markets, which is more rare, they're incredibly costly. Processed and fast foods tend to provide the most calories for the cheapest price, are always available, and don't spoil as quickly.

The Food and Agriculture Organization's flagship 2020 annual report indicated that healthy diets cost 60 percent more than diets that meet just the minimal nutrient requirements and 500 percent more than diets that meet just the energy requirements for basic bodily functions. These diets tend to be made up mainly of starchy staples.[3] Healthy diets are unaffordable for more than three billion people; 57 percent of populations throughout South Asia and sub-Saharan Africa are unable to afford them.

Even in the United States, while the average household in 2015 spent only 6.4 percent of their budget on food, the poorest 20 percent of households spent around 35 percent of their budget on food. At the same time, poor households in low-income countries spend 50 to 80 percent of their household budgets on food. In many places, the price of a basic diet that only meets caloric needs often exceeds daily wages.[4] Costs of living have continued to rise over past decades, but income levels have stagnated in comparison, so the struggle becomes more difficult. Economic downturns following crises like the COVID-19 pandemic make these struggles all the more challenging.

On quite a different level than individual poverty, trade policies significantly shape economic and nutrition outcomes. Increasingly, globalization tends to enhance the variety and availability of high-quality foods in larger population centers. However, trade has also been linked to rising levels of income inequality as well as the infiltration of cheap and unhealthy foods that have "empty calories" but are convenient choices.[5] Trade, while diversifying the global food supply, has other ramifications beyond the diets of consumers. It can increase competition and tends to favor producers who are able to offer goods at the lowest prices. Both of these factors generally reduce food prices for consumers even as they force producers to increasingly cut costs to make a profit. Since the 1980s, policies promoting global trade have increased the amount of food exported from low-income countries. In some cases, this has reduced the food available for these nations' own populations. At the same time, reducing levels of food exports may adversely affect the incomes of farmers in poorer countries.

Climate can also affect food prices and volatility. Extreme weather events such as droughts and floods (increasingly, climate change related) and unpredictable seasonal changes can be a trigger for major food crises and food price spikes. The most vulnerable and poor populations struggle to be resilient under these climatic conditions. With increased food prices, those with little income or resources will switch to cheaper foods such as rice or wheat to fill their stomachs. This can lead to more extreme coping strategies to stave off hunger and malnutrition. A study that examined the effect of seasonality of price volatility in Ghana, Namibia, Malawi, and Ethiopia found a causal link between maize prices and child malnutrition.[6] Between October 2004 and January 2005, maize prices doubled, resulting in a sevenfold rise in admissions to local health clinics for severe acute malnutrition.

The United Nations' report *The State of Food Security and Nutrition in the World* warns that "climate variability and exposure to more complex, frequent, and intense climate extremes are threatening to erode and even reverse the gains made in ending hunger and malnutrition."[7] The result could be political instability, violent conflict, deteriorating environmental stewardship, migration and forced displacement, and reduced control over food systems.

Price volatility and spikes can also be caused by conflict, and usually the poorest suffer the consequences of war. There are still a number of countries that are struggling with political instability, social unrest, war, and humanitarian crises, which can destabilize food systems and ultimately harm health in both immediate and longer-term ways. Protracted crises can create dysfunctional institutions, competition for natural resources, and inadequate access to health and social services. Conflict can both lead to and result from food insecurity and malnutrition.

The FAO classifies most countries currently experiencing conflict as having low incomes and food deficits, with high burdens of undernourishment and high rates of stunting.[8] Environmental scarcities, natural resource constraints, and food insecurity do not always result in conflict, but they can cause tense situations that can escalate into violence. In arid regions, like the Horn of Africa and the Middle East, access to water and land are significant contributors to conflict. In addition, countries vulnerable to volatility in the price of imported foods may suffer social unrest and conflict when prices rise and governments are unable to intervene.

Sometimes, assailants use hunger as a weapon to force populations into submission. A report released by the United Nations highlighted hunger as a war crime committed against civilians in Yemen.[9] In 2017, Yemen closed its ports, preventing vital humanitarian aid from reaching conflict-hit populations and requiring food to travel long distances to reach those in need. Targets of civilian infrastructure attacks included water facilities, food transportation systems, farms, and marketplaces. Fighting in Yemen's port city of Hodeidah, which is a vital gateway into Yemen, further intensified difficulties in obtaining food, clean water, and medical supplies. The closure of Yemen's ports pushed an additional 3.2 million people into hunger and tripled the price of basic commodities. By 2018, 80 percent of Yemen's population needed humanitarian aid, with at least 8 million people living on the verge of famine.[10] The United

Nations has described the situation as the worst humanitarian crisis of our time.

INEQUITIES ACROSS FOOD ENVIRONMENTS

Food environments are the places where people make decisions about what to eat and what to buy. These places can be a farmers market, a corridor of vending machines, a restaurant, a school cafeteria, a high-end grocery, or a corner store. These diverse food environments have complex influences on our level of engagement and decisions linking to the food we eat, based on convenience, appearance, scent, physical placement, pricing, packaging, and even menu layout and design.

Food environments are often tied to inequities. Disadvantaged communities lack healthy, safe, and affordable food sources that would compose a healthy diet. Food environments are shaped by the geography, structure, and diversity of our cultures, and their variable quality underscores systemic inequities of society. Whether our locales are urban or rural has implications for the availability, accessibility, and quality of the food we choose, procure, and consume. Some spaces provide access to local and seasonal foods, while others do not. Some food environments are sophisticated and modern, offering a wide variety of highest-quality fresh foods in ample supply, while others can be rudimentary, with only basic provisions and a lot of junk food available. The latter are the food swamps mentioned in chapter 1.

People living in food deserts or swamps have few places to purchase their groceries and find healthy meals. Dollar stores have taken over grocery stores in much of rural America, but with a concentration of unhealthy foods and very little that's even perishable. Allison Aubrey, a reporter for National Public Radio, did an interesting piece for *CBS News* on dollar stores.[11] Aubrey identified this as "a struggle between main street and corporate America. Dollar stores are increasingly influencing where we shop and what we eat. They don't sell fresh fruits and vegetables or meat." Just like urban centers, rural areas struggle with the proliferation of food swamps, in which fast-food restaurants, junk food outlets, convenience stores, and liquor stores greatly outnumber healthy food options. These food swamps have an even greater impact on obesity than do food deserts.[12]

Joel Gittlesohn, a colleague at the Bloomberg School of Public Health at Johns Hopkins University, has been leading the way to try to improve food environments in Baltimore, a redlined city, in which many neighborhoods are considered high-risk areas for investors. Minorities, especially African Americans, reside in these areas. This has led to a de-investment in these neighborhoods, including their food environments. Gittlesohn and his team worked with small food stores in disadvantaged, ethnic minority neighborhoods in Baltimore to improve access to healthy foods by working with wholesalers, providing nutrition and food preparation education on social media and at the point of purchase. They've seen dietary diversity improve in these populations through these targeted interventions such as stocking fresh, healthy foods in corner stores, designing better menus, and working with wholesalers to create healthier supply chains.

High-income cities and neighborhoods typically have more grocery stores that sell fresh produce and healthy foods, green spaces, and bike pathways that promote physical activity. They also tend to have access to better health care, which results in higher life expectancies and overall quality of life. Seattle, Tokyo, Manhattan, London, and Washington, DC, are considered expensive cities to live in, and the demand for green space and high-quality food markets justifies the investment. However, even in these affluent cities, pockets of poverty remain.

Our food environments are shaped by industry and marketing. Food marketing and ads are everywhere, tempting us in all kinds of ways. They appear on rural billboards, in the previews at movie theaters, as sponsors to major sporting events, and signs on the side of skyscrapers. Marketing of unhealthy foods is particularly prevalent in middle- and high-income countries where disadvantaged, marginalized, and poor people find themselves in settings that drive poor food choices. Classic industry tactics shift to lower-income areas when high-income consumers take less interest in their products. This was seen with tobacco companies that moved into low-income countries as smoking declined in the United States and continues as food and beverage industries push soda and junk foods around the world.

Some argue that junk foods produced by multinational food and beverage corporations with powerful marketing strategies tend to displace traditional food system offerings and dietary patterns by placing healthier food choices at a disadvantage through their promotional efforts.[13] In response, the food industry justifies the development and advertising of such products by claiming that they are merely "giving consumers what they want," which places the burden of making a healthy choice solely on the individual. Marion Nestle, a professor at New York University and author of the books *Food Politics* and *Soda Politics*, is a leading voice in calling out food industry tactics. In her words, "I could see that nobody was talking about food industry marketing and lobbying practices as factors in childhood obesity, but I could, and with impunity. It's gotten easier. Young people today recognize how corporations control food systems and how profits drive corporate practices. They hold the future, and I want them to have the tools they need to advocate for healthier and more sustainable foods systems."[14]

Efforts to advertise and market junk foods are particularly problematic when directed toward children, whether it's candy placed at checkout counters, colorful and engaging advertisements shown in-between cartoons on television, or unhealthy foods connected to games, toys, prizes, and merchandise. Children have few ways to defend themselves against these messages while also having the greatest need of adequate nutrition for optimal growth and development.

Despite differences in how poverty is experienced, the way food environments are designed puts people at a disadvantage and can be incredibly disempowering. People who are food insecure don't have the luxury of basing their decisions on health and sustainability, and they often don't have the range of choices of the more affluent. This creates further marginalization.

INEQUITIES IN MEAT PRODUCTION AND CONSUMPTION

Meat production and consumption raise some of the most contentious ethical issues associated with food systems. Is it ethical to feed staple crops to livestock animals to keep up with increasing demand for meat when there are still so many people who are going hungry? Is it ethical to kill and otherwise use animals for the sole purpose of food? Is it right to ask those who are undernourished, who get very little animal-source foods to begin with (and need them for their health), to limit their meat consumption along with high-income country consumers?

The value of animal-source foods for human health is hotly debated. However, these foods contribute substantially to human growth and development. Currently, meat, aquaculture, eggs, and dairy provide 18 percent of all calories and 37 percent of all protein consumed worldwide.[15] Animal-source foods supply multiple bioavailable nutrients missing in the cereal-dominated diets, and they're some of the best sources of high-quality, nutrient-rich food for children ages 6 to 23 months. Inadequate intake of nutrients readily available in animal-source foods, including vitamin A, vitamin B12, riboflavin, calcium, and iron, can cause anemia, poor growth, rickets, impaired cognitive performance, blindness, neuromuscular deficits, and eventually death. Micronutrient deficiencies have also been associated with brain-related disorders, including lower cognitive function, autism, depression, and dementia. While these nutrients are available in many plant-based foods, they're more heavily concentrated in meat, and they're more bioavailable, meaning they're easier to absorb.

But for many poor people, animal-source foods are cost-prohibitive. Nearly 800 million individuals consume insufficient amounts of these foods, with those who consume the least meat residing in Africa and South Asia, where populations are most afflicted with undernutrition.[16] In contrast, many people in high-income countries consume far more meat than they need for human health, and the amounts have been increasing. There are plenty of studies that show that those who are in good health can meet their nutritional needs and reduce their risk of noncommunicable diseases by eating a predominantly plant-based diet.

Average yearly per capita meat consumption (excluding fish and seafood) in the bottom four meat-consuming countries (Sudan, India, Bangladesh, and Ethiopia) is less than one-thirtieth that of the top four countries (Brazil, Uruguay, Australia, and the United States). The average person in Bangladesh consumes 3 kilograms of meat per person per year, compared with 124 kilograms per person per year in the United States.[17]

Demand for meat in middle- and high-income countries continues to increase, despite current overconsumption. Global production of poultry meat grew more than twelvefold between 1961 and 2014, while cattle meat production has more than doubled since 1961. Per capita meat (not including fish and seafood) consumption in China is roughly 15 times greater than it was in 1961.[18] Pigs are the most popular animal to consume, with China consuming two-thirds of all pork worldwide—more than 400 million pigs annually. To keep up with demand, global meat production has quadrupled since the 1960s.

Furthermore, as low-income countries gain economic wealth and stability, the consumption of animal-sourced foods is likely to increase, which will enable vulnerable populations to get a share of the nutrients that promote better health. Global demand for beef is projected to increase by 95 percent between 2006 and 2050, while demand for animal-based foods generally is expected to rise by 80 percent over the same time period.[19] The demand for poultry meat is expected to grow by 121 percent and the demand for eggs by 65 percent.[20]

The animal production systems that generate all this meat have many negative environmental impacts. They contribute to greenhouse gas emissions and pollution. They contaminate surface and groundwater. They encourage the conversion of land for grazing. Their effects on climate change are particularly pronounced because of the low efficiency of converting feed to meat, the fermentation of food by ruminants, and manure-related emissions. Furthermore, many of these negative environmental effects of meat production fall on those who cannot afford eggs, meat, and dairy products.

Animal production also contributes to a loss of biodiversity among livestock. Today, a few highly productive breeds adapted to industrial production systems have replaced most local breeds around the world. Between 2001 and 2007, 62 livestock breeds became extinct for myriad reasons. Of the 40 livestock species consumed by humans, 5 species account for 95 percent of today's meat production.[21]

The environmental effects of livestock production depend on the animals being raised, where they're raised, and the owners' production practices. The consumption of livestock products can be broken down by the type of animal (sheep, poultry, beef), the type of product (meat, dairy), the production system (organic, free range), and other factors. Over time, increasing numbers of animals have been raised in industrial settings where they can be strictly monitored and controlled. Antibiotics are given to livestock to reduce infections caused by underlying care, overcrowding, or sanitation issues, which raises concerns about increased antibiotic resistance in humans. Such production systems have been designed for efficiency, not for animal welfare.

Different meat production systems also have varying impacts on the environment. In Pakistan, Mongolia, Nigeria, and Mali, pastoralist systems still exist in which animals are raised through nomadic practices. At the other extreme, the United States is dominated by concentrated animal feed operations in which animals are raised in extremely confined and largely stationary settings. In Ireland, cows are sustained on grass-fed and pasture-raised systems. Different systems have widely varied effects on people's livelihoods as well as environmental sustainability.

Ruminants, or grazing animals such as beef cattle, have a much greater impact on the environment than do animals with higher feeding efficiencies, such as pigs and chicken. Most commonly, ruminants are raised on grain that people could eat, and this grain is grown on arable land that could be used to grow other crops. To satisfy the demand for meat, livestock animals currently consume 27 percent of the world's calories. Beef is particularly resource intensive, requiring up to 30 kilograms of grain to produce 1 kilogram of beef.[22] As opposed to other livestock animals, ruminants can graze on lands that have no alternative uses. Unfortunately, this is not the norm. Only 14 percent of cattle are raised on pasture, with another 16 percent raised on discarded crop by-products.

Including Meat in Your Diet?
Here Are Some Points to Consider

Whether you're a vegetarian, an omnivore, a flexitarian, a pescatarian, or a ketogenic eater, you're likely aware of both the environmental impact of raising livestock and the animal welfare issues associated with raising or catching and slaughtering animals. Beliefs about whether animals should be eaten and how they should be raised vary from culture to culture, but here are some points to consider as you define and refine your diet:

- **Individual health:** Humans don't need to consume a lot of animal protein to remain healthy. Eating too much meat leads to heart disease, cancer (from highly processed meats such as bacon), and other medical issues, whereas focusing on a plant-dominated diet has been shown to be beneficial to health.
- **Pet health:** Meat eaten by household cats and dogs is often overlooked by families that consider their overall animal-sourced foods consumption. Pet food is a huge industry, one subject to the same marketing ploys used for human food.
- **Environmental health:** Methane gas emitted by grazing animals, such as cows, damages the environment, contributing as much as 14 percent of all greenhouse gases. The destructive 2019–20 wildfires in the Amazon basin were prompted by clear-cutting forests to open up grazing land for cattle and soy, destroying untold numbers of flora and fauna in the process.
- **Global health:** Zoonotic diseases such as avian bird flu, swine flu, and COVID-19 are due to spillover events from other animals

(such as bats) to humans. COVID-19-related outbreaks in meat processing plants in the US also resulted in infected plant workers and the "depopulation" deaths of millions of herd animals that never made it to markets.

Making decisions about what to buy that's sustainable, humane, and healthy can be overwhelming; here are a few things to look for:

- Seafood Watch, the Monterey Bay Aquarium's program for sustainable seafood identification, helps consumers and businesses choose less environmentally damaging, marine-harvested foods.
- For meat, look for free-range and organic labels. Companies that don't overuse antibiotics will advertise that on their labels. There are also third-party animal welfare verification labels, including Humane Farm Animal Care (Certified Humane), Animal Welfare Approved, and Food Alliance Certified.
- Deciphering egg carton labels can be tricky. Only the "Certified Organic" label means some animal welfare claims are verified by the US Department of Agriculture for example. Otherwise, look for "Free Range" or "Pasture Raised" to select eggs from uncaged chickens that have outdoor access ("cage free" does not mean they have outdoor access).
- For anything you can't find on packaging, ask your butcher or server who supplies their meat and seafood, then visit the companies' websites and look for their claims regarding animal welfare standards—but beware of misleading marketing language.

As with other food systems, animal production systems both contribute to and are significantly affected by climate. Rising temperatures and other stresses can cause animals to produce less milk or to grow more slowly, producing less meat and decreasing the calories and nutrition available for human consumption. Changes in precipitation can lead to water scarcity, which can cause animal dehydration. Large stresses can lead to losses of animals that can be severely detrimental to ranchers' livelihoods. Over the past two decades, in Africa south of the Sahara, cattle losses have been as high as 20 to 60 percent of the total population during periods of severe drought.[23]

Despite their environmental impacts, animal-source foods are an important part of people's identities. My team at Johns Hopkins University has done some work in Northeastern Kenya with pastoralists who roam with their animals. These herders value and cherish their animals, and they're a huge part of their tradition, culture, and livelihoods. However, the effects of climate change and resource scarcity are increasingly threatening these populations. In interviews my team has done with nomadic herders from the Borana ethnic group, one female herder said, "Pastoralists are faced with the hard task of grazing their livestock because they're caught up in the middle of human settlement and farms. If you graze on farms, this will lead to conflict. There is also insecurity since there are frequent raids from neighboring communities due to drought and famine. Herding of livestock poses risks of loss of life due to the raid. One can easily be killed while herding."[24] Food, water, and land access are all becoming more and more constrained. Already, there are decreased numbers of those who choose pastoralism as a livelihood because of climate change. Unfortunately, this choice has some negative repercussions. Research has shown that when pastoralists no longer roam and instead settle down in one location, their health status typically declines.[25]

In 2019, my team and I also researched why Americans value beef so much, in part by conducting interviews with consumers and producers in the United States to better understand their lifestyles and beliefs. When we spoke with ranchers, we discovered that they consider beef production to be part of their culture, identity, and heritage. Ranching is a way of life that defines these people. Many even voiced concerns about their mental health if they were suddenly unable to maintain their livelihoods. One 65-year-old small-scale production operator said, "If it wasn't for the way of life, and the peace of mind, and the mental therapy that you get out of owning the place, okay, the returns are low enough that you either have to be passionate about doing it or you had better find something else to do."[26]

During this project, we also made the heartening discovery that many producers and ranchers are adopting more sustainable practices. One 47-year-old Nebraskan who worked for a large-scale production operation said,

> Sustainable can be somewhat of a buzzword that's overused, but we define that as it applies to people, community, natural resource conservation, and profitability. Those are our goals. To be able to have success in all of those areas simultaneously, with success in one not at the expense of success in another. For instance, I don't want to be profitable at the consequence of poor natural resource management. We don't want to put profits ahead of having good people and taking care of people in the business, or to have our communities in decline.[27]

Animals have been integral to human life since the beginning of human existence. The Industrial Revolution brought changes in the way animals were viewed, with an increased focus on their consumption. Many animals consumed globally are raised in concentrated animal feed operations in which animal welfare principles are considered only for industrial production purposes and for food safety of humans without much regard for their intrinsic well-being.

The livestock sector is currently the world's fastest-growing agricultural subsector. On average, livestock production accounts for 40 percent of the global agricultural gross domestic product in developing countries. More than half of the world's poor people rely on the livestock sector for sustenance, income, insurance, and food. Thus, livestock production will likely play a prominent role in efforts to eliminate poverty.

If current best sustainability practices were used for all live-stock production, livestock greenhouse gas emissions could be decreased between 14 and 41 percent.[28] Other practices that could reduce the environmental footprint of livestock production include refocusing on animal-source foods other than beef cattle and using mixed agriculture systems in which farmers simultaneously grow crops and raise livestock. Cattle can also be raised with increased feed conversion efficiency. Ranchers in the United Kingdom, for example, have moved toward more efficient cattle breeds and pasture systems to decrease the number of animals but maintain production, resulting in a 28 percent reduction in methane emissions over that period.[29]

Sustainable practices will require more efficient use of resources paired with environmental stewardship. The adverse environmental effects caused by livestock production can be reduced by such strategies as improving herd efficiency, health, and genetics; improving feed production and feeding practices, including grazing management; reducing herd sizes to retain only productive and efficient animals; ensuring attainment of market size or weight earlier; and managing manure to recover and recycle nutrients and energy. Massive industrial operations must be curbed while more sustainable, humane, and balanced practices are promoted.

Increased sustainability for animal-source foods will also require changes on the consumer side. Those who consume too much meat will have to reduce their intake, while those who consume too little meat will need greater access to animal-source foods. There are recommendations for high-income countries to reduce consumption of animal-source foods and for growing economies to avoid high consumption. The World Resource Institute calculates that adults would need to decrease their consumption of animal foods by 30 percent to meet greenhouse gas emission targets set out by the Paris Climate Agreement. Many ethicists argue that if certain resource-intensive foods, such as animal meat and by-products like dairy, are considered vital for human health, they should be equitably supplied to entire populations.

Low-resource alternatives to livestock should also be considered to combat micronutrient deficiencies and as a protein alternative. Although foods such as mollusks and insects (which are regular parts of diets in some parts of the world—in Thailand, it's water bugs; in Mexico, it's chapulines [grasshoppers]; and in Africa, locusts and termites) may be less popular, they can significantly improve nutrition outcomes while causing less environmental degradation than the production of other, larger animal-source foods.

There are major concerns with the expansion of industrialized animal systems and demands for meat because they're designed for efficiency, not ethical treatment. Humane alternatives could be considered to fill nutrient gaps for all countries that require less resources. These foods, such as farmed fish, mollusks, insects, and protein-rich plant food, make significant contributions to nutrition, leave a smaller footprint on the planet, and are less harmful to animals.

Technological innovations could transform livestock production practices, but big questions remain. Plant-based meat alternatives and cultured meats will become increasingly important in the future, but will these products be acceptable in taste, widely accessible, and sold at the right price point? Will they remain a highly processed, nutritionally questionable alternative? What energy resources will be required to produce these new foods?

Low-income countries are still catching up to the rest of the world regarding livestock production, and many people in places such as Africa argue that now is their turn: high-income countries caused the problems these countries face while developing countries have never had a chance to produce or eat animal-source foods. Unfortunately, in the context of climate change, the situation has changed. It may not be anyone's turn if we continue with the status quo when it comes time to address climate change. We need to build back inclusive, sustainable, and healthy food systems for everyone. Some will have to make more sacrifices, particularly those with abundant resources consuming more than they need.

Do animal-source foods support or harm sustainability and health outcomes? In reality, they do a bit of both. Climate change is not the only measure of sustainability. Sustainability also describes human and animal health outcomes and well-being, equity, and security. Often, discussions about the sustainability of animal-source foods neglect to include the effect that low consumption of animal-source foods has on the lives and futures of nutritionally vulnerable populations, women, and children. Moving forward, we'll have to combine more sustainable livestock production practices with increased access and moderate consumption. Clearly, those living in high-income countries could come down on the amount of animal-source foods they consume. It's just not necessary to consume animal products three times a day, every day. These measures could improve nutrition, livelihoods, food security, and health while reducing the environmental effects and alleviate the animal welfare concerns of livestock operations.

INEQUITIES FOR WOMEN

Women keep food systems functioning and moving because they make up the majority of workers in the food sector and most have a second job—to feed their families. Their role is critical throughout the food supply chain, from production on the family plot to working lines in meat production plants to shopping to food preparation within the household. Yet they're often ignored, forgotten, or disenfranchised, especially in circumstances where mothers find themselves unavoidably involved in what can often be the drudgery of smallholder, subsistence farming.

Women carry a heavy burden in many parts of the world. In northern Ethiopia for example—considered the breadbasket of the country for its role in growing the teff used for injera, the nation's ubiquitous spongy bread—girls can be married off very young, between 13 and 14 years old. Right away, they're expected to start having children. On top of that heavy burden, they're responsible for fetching the water for their household, often walking 15 to 20 miles a day and returning home carrying 50-pound water jugs on their heads. Women are expected to care for their children and husbands and often are responsible for other in-laws and relatives. Women are still dying in childbirth in many parts of the world but particularly in low-income countries with strained health systems. Having spent time in these places, I find it astounding that these women, who are responsible for so much and are heavily relied on, are able to stay healthy and functional given their lack of agency and opportunities and their secondary status compared with men.

In addition to caring for their families, women are often responsible for producing and acquiring food, although their contribution to these efforts is often overlooked. A report from the FAO found that women account for 60 to 90 percent of food production globally, with women in developing countries making up approximately 40 percent of the agricultural labor force.[30] Most smallholder farmers are women, and as urbanization creates job opportunities in nearby cities, women may be responsible for an even greater share of the burdens of food production and maintaining the household. Because of the out-migration of men to cities or to other countries, women also may be responsible for conservation efforts to protect against biodiversity loss, as they are often the ones who have knowledge of traditional varieties and practices that can contribute to improved health and environmental outcomes. You see this in places like Nepal, where most of the men are building skyscrapers in Dubai or World Cup soccer stadiums in Qatar. The women stay behind and become the main caretakers of their land along with their families. Some have called this phenomenon the "feminization of agriculture."

Even as women take on more tasks in agriculture, they continue to have limited access to and control over land and family finances. Far fewer women than men hold ownership of livestock and land, and those who do have smaller plots than their male counterparts. The FAO estimates that a 12 to 17 percent decline in undernourishment could be achieved by addressing the gender gap in agriculture.

Women are disproportionately bearing the ill effects of climate change. Climate change is harming maternal and child health through an increase in infectious diseases and a decrease in food intake due to less food availability and higher food prices. With increasing temperatures and heat stress, the rate of preterm births may increase in populations where women participate in agricultural work, and bending over and standing all day long appear to exacerbate this risk.[31]

In many regions, enforced gender roles and norms deprive women of equal access to education, land rights, or financial services. Women who are employed tend to earn lower wages than men. A substantial body of literature indicates that improvements in women's socioeconomic status have long-term benefits for child and household nutritional status, health care, and education attainment.[32] When women are better educated and able to access resources to improve food security, their health and the health of their children greatly improve.[33] In particular, the age of marriage, the age of first pregnancy, the spacing of pregnancies, and the number of children they have all have significant effects on maternal and child nutrition.

Gender equity is a priority in the food sovereignty movement, which asserts that the people who produce, distribute, and consume food should also control the mechanisms and policies of food production and distribution. The movement argues for a redirection of power away from large companies and toward the people, and in agriculture, these people are often women.

Every decision made in food systems usually means someone or something will win and another will lose. While governments should weigh the pros and cons in their attempts to create policies that incentivize the production and consumption of healthy foods for everyone, they often do not. The food and beverage industry is under increasing pressure to produce and market healthier, sustainable foods, though they often shift this responsibility onto the shoulders of consumers and target those with little voice or power. Consumers face many trade-offs in their dietary choices often shaped by price, where they live, who they are, and society's social norms.

It's critical to bend food systems toward those who provide healthy food in equitable ways to the world's population without causing irreversible damage to our planet. This is not that hard to do as long as there is political will by governments, strong incentive structures for businesses, consumer awareness and willingness, and community empowerment. We need to create more accountability within governments and the food industry to address the inequities that exist within food systems and ensure that there's a viable planet for the next generation. The next two chapters will focus on how we can, collectively, make this happen.

CHAPTER 4

Can Better Policies Create Better Food?

TRANSFORMING THE GLOBAL FOOD SYSTEM will require changes at two broad levels: from policy changes (discussed in this chapter) and individuals' actions (discussed in chapter 5). The agenda is immense. Food policies should help ensure that all people have access to safe, healthy, affordable food; that farmers and workers are supported; that animals are treated humanely, and that air, water, and land are protected for future generations. Current food policies in the United States — or anywhere else — don't achieve all of these goals. On the contrary, not a single nation has a holistic food system policy designed to improve human nutrition and well-being while protecting the environment. Many countries have an agriculture policy, dietary guidelines, or even climate change policies, but very few if any bring those policies together in a coherent, all-encompassing strategy that addresses the entirety of food systems, and their goals can sometimes contradict each other.[1]

Globally, agriculture policies tend to be focused on improving yields of the basic staple grains (soy, maize, rice, wheat), subsidies for the production of those staples, and uniform production processes. This paradigm has many flaws. It lacks an effective integration of health and environmental issues. It overlooks the need to prevent food loss and waste. It fails to promote healthy diets that can be accessed by everyone. And it's ineffective in addressing disruptions to food systems caused by conflicts, climate-related natural disasters, and pandemics.

In fostering healthier diets, governments tend to favor interventions focused on individual-level actions and initiatives. But food choice is not simply a personal decision and these interventions can actually worsen dietary inequities. Diets are shaped by where you live, who you are and what options you have, and are also driven by deep (often unseen) systemic social factors and injustices. Approaches that focus on population health policy rather than on those that require personal agency are more effective, equitable, and enforceable.

Food policies affect all steps of the food supply chain, from production through processing, distribution, marketing, purchase, and consumption. The policies involved include those affecting agriculture, health, nutrition, dietary guidance, the environment, water, food waste, bioenergy production, trade, transportation, and economics. Food policies reflect the complexities of food systems. Fortunately, these complexities offer abundant opportunities for change.

With the right package of interventions and appropriate implementation, improvements in health and environmental resilience can occur rapidly. Government policies are often contradictory, such as the misalignment of agriculture subsidies for not necessarily healthy staple and oil commodities as compared to the promotion of diverse, healthy diets through food-based dietary guidelines. Governments need to coordinate and institute multiple interventions, implement them simultaneously, ensure they're complimentary, and prioritize the needs of their citizens over other vested interests.

IMPROVING DIVERSITY OF FARMING SYSTEMS

Today, the main challenge for the agriculture sector is to simultaneously provide enough food—in both quantity and quality—to meet everyone's nutritional needs while conserving the natural resources to produce food for present and future generations. Policies should be geared simultaneously toward human and planetary health. There's no one-size-fits-all solution; many different approaches can be taken, depending on the food system and population, the climate, and the ecosystems in which foods are grown.

One of the most important functions that governments can do to promote health and sustainability within food systems is to uphold agricultural diversity. Studies have found that diversified agricultural systems have 20 to 60 percent higher yields than monocultures in some specific contexts.[2] In addition, increased agricultural diversity on farms can potentially improve diets at the household and local levels by making nutrient-rich foods available throughout the year. Optimizing biodiversity on landscapes and managing water and other natural resources can build healthy agroecosystems and secure livelihoods.

Ministries of agriculture can enhance agricultural biodiversity and nutrition by increasing access to seed varieties and livestock breeds that are diverse and therefore resilient to weather conditions, pests, and diseases. The use of cover crops, crop rotation, manure, and appropriately applied fertilizers can improve soil quality and potentially enhance the nutritional content of foods.

Governments need to support farmers groups, community-based organizations, and social movements that encourage diversification and provide strong agriculture extension support. An example of such support was a large-scale, interagency, community-based approach in Pohnpei, Micronesia, to promote biodiversity and address the shift away from traditional diets. The initiative, started by the late Lois Englberger in 1998, a passionate local food advocate and nutritionist, developed two slogans: "Go Yellow" promoted yellow-fleshed varieties of plants and "Let's Go Local" more broadly promoted the production and consumption of the vast variety of local foods as opposed to monoculture commodities imported from other countries.[3] Support for these efforts included workshops, demonstrations, school visits, youth clubs, breastfeeding clubs, mass media interventions, and promotional materials. This intervention also led to a set of guidelines that other nations can use to support local biodiversity and better nutritional outcomes.

Gaining public support for such changes won't be easy. In Ethiopia, Uganda, and Kenya for example, where the terrain and weather conditions can be unpredictable, shifting toward more diverse crops can be risky so subsistence farmers are skeptical of transforming their agricultural practices to grow different foods. When discussing diversification of landscapes and producing a variety of crops from those landscapes, farmers often respond, *We can't even produce enough food. Why are you talking to us about diversifying the types of food we are growing beyond our staple and cash crops?* Farmers don't want to compromise their core source of income and divert resources to grow other crops with uncertain profitability or where there's no guaranteed market. This is the constant tug and trade-off between nutrition and agriculture, and it extends all the way to the highest levels of government. How do you make nutrition demand-driven and how do you incentivize farmers and the systems they work within to think of nutrition as a goal along with generating income?

Many farmers offer examples from which others can learn. The Food and Agriculture Organization (FAO) often highlights globally important agriculture heritage systems as "outstanding landscapes of aesthetic beauty that combine agricultural biodiversity, resilient ecosystems and a valuable cultural heritage. They sustainably provide multiple goods and services, food, and livelihood security for millions of small-scale farmers and indigenous peoples."[4] The FAO argues that these systems embody important agricultural practices, such as low water use, improved soil techniques, protection of natural resources, and high levels of biodiversity, all of which contribute to food security and diets.

Globally important agriculture heritage food systems offer valuable lessons on sustainability. Indigenous peoples understand that their local foods are resilient and adapted to their environments. They know the animals and plants that make up the natural resources in the world's forests, pastures, rivers, lakes, and seas and how it all fits together in their ecosystems. Their knowledge of these resources is grounded in their cultural and historical legacy—these plants and animals, and the foods made from them, connect them to their ancestors. While learning from these systems could provide governments with ways in which agriculture production policies can be created and implemented to better serve the planet, there are trade-offs. Some places don't have a history of heritage food systems, or the indigenous peoples who curate those lands have been marginalized and rendered powerless. Unfortunately, governments focused on generating enough calories to feed a growing population often de-prioritize the other benefits that land can bring to local communities.

Ministries of agriculture should strengthen and invest in their agricultural extension programs and the agents (community-based agricultural specialists) who provide information, training, and tools to food producers. But publicly funded extension services around the world are in decline, and their ability to expand beyond a certain set of crops or farm services is limited. The typical training these agents provide to farmers includes advice on how to grow a narrow set of crops, apply pesticides or herbicides, and undertake crop rotation.

In partnership with the World Bank, my team examined whether extension agents around the world are able to integrate nutrition into these basic services. We found that there's little integration into their current work packages. Challenges included a lack of training for agents on how to integrate nutrition into their current agriculture services, unclear mandates from the ministries of agriculture, little in the way of providing transportation or phones to get out into communities, and disempowerment of women who work in extension services. There were also more significant, systemic challenges between agriculture and health sectors in that nutrition falls between the cracks with no one taking responsibility.[5] What it came down to was insufficient training to build a solid workforce that had sophisticated skills to help farmers in challenging times.

Some places in the world have a dearth of human workforce in food systems, sometimes due to long-term conflict. Glenn Denning, a professor at Columbia University and the person who taught me so much about agriculture, worked in Cambodia with the International Rice Research Institute in the 1980s to rebuild their rice crop systems and workforce following the genocide by the Khmer Rouge. During the early days when he worked there, Cambodia's Hun Sen government was only recognized by the Soviet Union and Soviet Bloc countries, including Cuba, as well as one other country not in that sphere of influence—India. Denning said, "I think the general principle when you come out of conflict is you often have minimal infrastructure, and very limited human resources, and you often need to borrow technology and build up research capacity yourself."[6]

With minimal access to extension services and low capacity, it will be difficult for some farmers to take on or adapt to new technologies. A woman smallholder farmer in rural Rwanda who has minimal tools and zero mechanization at her fingertips could be left behind as more complex technologies come online. However, these technologies could be game changers, allowing her to leapfrog over decades of past technologies to more efficient ways to farm. Technologies such as precision agriculture, which incorporates GPS (global positioning system), drones, robotics, soil spectroscopy, geospatial mapping, and cloud computing all help to manage fields more efficiently.[7] Innovative technologies will all require funding of research and development at universities and tech companies, along with political support and regulation. While recognizing the many different perspectives regarding new technologies, governments will need to streamline regulatory processes to prevent massive barriers and delays to implementation.

FARMING WITH NUTRITION IN MIND

Nutrition-sensitive agriculture is another approach to sustainable farming and cropping systems. Nutrition-sensitive agriculture aims to sustainably intensify food production to increase the nutritional quality of crops, not just high yields. Currently, 36 percent of the calories produced by the world's crops are being used for animal feed, and only 12 percent of those feed calories ultimately contribute to the human diet (as meat and other animal products).[8]

In Timor-Leste, where approximately 60 percent of children under age 5 are chronically malnourished and almost 39 percent suffer from anemia, I worked with a program called Seeds of Life within the Ministry of Agriculture to develop a nutrition-sensitive agriculture strategy that would also improve food security. My particular goal was to advocate for more investments in nutrition-sensitive agriculture beyond staple crops such as rice and cassava. The challenges I faced illustrate some of the complications in implementing this approach. It was difficult to convince the Ministry of Agriculture to think about agriculture from a nutrition perspective, because they're primarily concerned with increasing productivity of rice for income generation. Nevertheless, Seeds of Life successfully implemented several important programs, such as introducing new types of crops—legumes and nuts suited to the local environment.

Another form of nutrition-sensitive agriculture is to invest in integrated and holistic land use and tenure policies. Mixed crop and livestock systems can minimize the impacts of livestock on climate and improve the nutritional quality of food. Such systems offer resilience to crop losses while providing additional income that's more stable than the income from crops or livestock alone. Additional examples include rice-fish aquaculture systems, poultry-orchard systems, and livestock–cover crop systems.

In Malawi, intercropping of maize with legumes has improved soil health and dietary diversity and has contributed to nutrition education at the community level.[9] These changes have also contributed to significant improvements in weight for age in children under 5. In Bali, the beautiful rice terracing system that integrates aquatic life, ducks, and rice provides a harmonious ecosystem. But the question remains whether these are scalable and cost-effective practices for governments to take on.

One of the best examples of ecological complementarity is the Mesoamerican "three sisters." The combination of corn (a grass), beans (a nitrogen-fixing legume), and squash (a low-lying creeper) maximizes their growth and optimizes their nutrition. Corn maximizes photosynthesis because it is a grass that grows tall and straight. Beans use the cornstalks to climb toward the sun, and they fertilize the soil to promote the corn's growth. The squash stays near the ground and shades the soil to retain moisture with its broad leaves. From a nutrition point of view, corn is an important source of carbohydrates and protein; beans are also high in protein, iron, and B vitamins; and squash can add vitamin A and fiber.

Policies involving aquaculture will become even more important to health and sustainability.[10] Moving aquaculture offshore or to land-based recirculating systems can reduce environmental footprints along with use of renewable energy sources in the aquaculture systems. Multitrophic systems that farm shellfish and seaweed along with larger fish such as salmon or trout can promote a more sustainable feeding system. Although aquaculture is becoming increasingly efficient, strategies need to focus on lower-intensity species that can be raised on plant-based proteins and oils, such as tilapia, catfish, and carp, as well as bivalve mollusks, such as mussels and clams.

Integrated aquaculture systems will also be needed to counter climate change. Rice-fish farming, a practice that emerged more than 1,700 years ago in China, and now in use in Cambodia, Bali, and Bangladesh, simultaneously produces rice, which is crucial for food security, and fish, which provides a valuable source of high-quality protein, essential fatty acids, and important micronutrients. Although rice yields tend to be lower in rice-fish farming than in intensive systems, the contributions to nutrition overall are far greater. These systems also enhance environmental sustainability, and the fish and other aquatic organisms raised in these systems provide pest management and natural fertilization.[11] Questions do remain, however, about their time demands and work-intensiveness.

Government support of organic farming can reduce the environmental impact of agricultural practices. Support can be in the form of research, competitive grant programs, and agriculture extension. Organic farmers often grow more crops and varieties, increasing biodiversity on the landscapes in which they work. Organic farmers generally exercise more sustainable practices, such as the use of compost, nitrogen-fixing crops, cover crops, crop rotation, and no-till practices. Their methods also decrease pesticide exposure for farmworkers and those living in surrounding areas. These practices result in more biodiversity, richer soil, less erosion, and less water and air pollution. The jury is still out on whether growing food using organic practices provides more nutrient-dense products.

But still, organic produce is substantially more expensive than conventional produce.[12] As a result, not everyone can afford to eat all, or even some, organic food. As the organic food market grows, prices are starting to drop because of the sheer volume of products, less-restrictive government policies on what is considered organic, and crop insurance programs that support small organic farms. Fresh, organic foods have benefits for human and planetary health and well-being, but if they aren't widely accessible and affordable, they can contribute only so much to mitigating climate change and diet-related diseases.

REJIGGERING THE SUPPLY CHAIN
FOR FOOD SECURITY AND NUTRITION

———

Greater governmental support for safe storage, processing, and preservation techniques could help ensure that people have access to safe food. In food systems not yet highly modernized, strategies should focus on improvements to storage and transport infrastructure, such as cold chains that maintain perishable food temperatures to ensure their safety from harvest to consumption. In more modernized food systems, innovative and sustainable technologies for storage and distribution should be implemented and their effectiveness studied. Satellite technologies, including GPS, have recently emerged to allow shippers and carriers to monitor the quality of their cargo and to shorten the cargo delivery time. These practices could not only increase profit but also reduce spoilage and improve food safety.

Ministries of agriculture should encourage or require processers to experiment with new techniques to preserve the nutritional quality of foods and reduce added salt, sugar, and unhealthy fats. Such techniques include fermentation, drying, and food product reformulation and fortification. Policies could be enacted that require processers to preserve or add micronutrients into foods during processing or to remove less healthy ingredients. Support for processing practices such as canning, freezing, and aseptic packaging technology can also lead to longer food shelf lives and ensure that these foods reach vulnerable populations who lack access to fresh and nutrient-dense foods.

Supporting a diverse set of food supply chains can increase the resilience of food systems. The COVID-19 pandemic sparked debates because supply chains were disrupted causing food shortages and insecurity. Short supply chains and alternative retail infrastructures can provide viable, accessible, and affordable alternatives to mass retail outlets that may be hard to reach for some consumers. Networks and micro-hubs of food producers could increase market access and limit food loss, and governments and large corporate entities (like Walmart and other major grocery store chains) can support local food by repurposing infrastructure in cities to favor farmers markets, mobile food trucks, and community food centers.

In the past few years, several US states have created policies that transport produce from local farms to school cafeterias, encourage farmers markets to accept food stamps, and provide greater opportunities for small and midsize farmers to sell their products locally. According to data from the US Department of Agriculture, more than 80 percent of America's 8,600 farmers markets in cities, suburbs, and more rural areas accept Supplemental Nutrition Assistance Program (SNAP, formerly known as Food Stamps) payments, a step in the right direction for equity.[13] In Brazil, school meal programs are being linked to local farmers to provide their produce to schools at lower cost.

Governments and the retail sector can enforce the use of packaging strategies that reduce and eliminate food and packaging waste. Many areas in the United States and countries such as Kenya, Mali, Cameroon, Morocco, and Rwanda have banned or taxed plastic bags and straws. These types of measures should be expanded to promote more sustainable, reusable options.

In some places, producers do not have much incentive to grow perishable foods, such as fruits and vegetables. These foods are often not purchased by wholesalers because of odd shapes or markings, dents, or impurities on their skin or surface. People need to be more accepting of "ugly foods" that taste just the same as the prettier ones. Some companies are attempting to change consumer opinions about ugly food and make these foods available at a lower price. Companies such as Imperfect Foods sell discounted ugly fruits and vegetables to consumers that would otherwise be discarded due their unappealing shape. In France, the supermarket Intermarché launched a program "Inglorious Fruits and Vegetables" in which they discounted disfigured produce 30 percent. It worked. Sales are up.

Other companies are transforming "unsellable" foods into juices or other acceptable forms for consumers. Companies are increasingly taking advantage of edible foods typically discarded in an effort to increase the world's food supply, and these practices can be scaled up. In some countries, the private sector has sought to package foods in ways that make it more convenient for people to cook and eat healthy meals. The City of Oslo has

implemented a new food waste bag system to recycle household food waste and other cities—such as Rome, my home in
recent years—have similar composting programs. Storage and
processing can also help minimize food loss. In the Philippines,
airtight, reusable "super bags" designed by the International
Rice Research Institute helped reduce rice crop losses to air, water, insects, and rats by 15 percent. In some places, such as West
Africa, solar drying can preserve the shelf life of perishable fruits.

CHANGING THE CHOICE ARCHITECTURE
OF FOOD ENVIRONMENTS

Changing food environments to promote healthier, sustainable
food choices converges on policy areas where action is needed,
including nudges and "choice architecture" (the way environments are designed to influence consumer decisions, such as
what products are at eye level, or positioned at check-out lanes);
nutrition labeling; food provisioning in specific settings, such as
schools; economic incentives and disincentives, such as retail
subsidies and taxes; and food promotion, including advertising
and marketing. Limiting unhealthy foods in these environments
will lead to greater integration of markets with communities,
which could strengthen people's connection to the food they
eat and where that food comes from.

Governments can also improve food environments by implementing policies that encourage supermarkets to supply

nutritious foods at affordable prices. South Africa has taken this approach; private health insurance companies have part- nered with supermarkets to improve buying behaviors, which have helped to increase nutritious food purchases and lower the consumption of foods high in salt and/or sugar, fried foods, processed meats, and fast foods.[14]

When it comes to marketing and retail, many food environ- ments could be better designed, using more effective choice architecture to influence good dietary decisions (rather than mislead or coerce consumers into poor decisions, which is often the case). Choice architecture is a way to design food environments to ensure that healthy food choices are easy to see, to order, and to choose from; are attractive in their price and appearance; and are easy to serve and eat. By changing the choice architecture, an environment can influence people's decision making, for better or worse. Primary to the design are "nudges," defined by Nobel Prize laureates Richard Tahler and Cass Sunstein as "any aspect of choice architecture that alters people's behavior in predictable ways without restricting any options or significantly changing their economic incentives such as time or money."[15] Nudging favors individual decision making over regulation or restricting choices. They are subtle ways to persuade in the places where people live, shop, work, and learn.

There are many examples of nudges that can be tried in food environments to encourage healthy eating. Fast-food chains should reconsider default side servings, such as including fries with a meal. Why not make salads the default? The Cool Food Pledge, initiated by the World Resources Institute, is setting an example by partnering with institutions like hospitals, corporations, municipalities, universities, and large retail outlets (Bloomberg, Harvard University, IKEA, Morgan Stanley, the World Bank Group) that represent a combined 800+ million meals served annually to their staffs, citizens, and customers. These entities commit to menus with "delicious climate action" selections as well as to the guidance and metrics tracking components. Signatories to the pledge will collectively reduce greenhouse gas emissions by 25 percent by 2030; many see this as part of their overall sustainability missions. Similarly, in cafeterias, trays and plate sizes could be smaller. In work canteens, the salad bar should be sitting at the center, making it harder to overlook. Not all nudges work, of course. Labeling menu items at restaurants with information about calorie, fat, sugar, and salt content has shown little effect. Just Salad, a fastfood salad chain, has added carbon footprint metrics to their menus. Although this is a step forward, most people don't fully understand recommended carbon levels (Is my chicken salad with 0.14 kilograms of carbon dioxide emissions bad for the planet?) or may become overwhelmed. Other healthy nudges include making healthier foods more visible and prominent than unhealthy foods in supermarkets and ensuring they can be quickly packaged for takeaway. Portion or package size of unhealthy foods could be reduced while increasing the serving size of healthy foods (all wrapped up in sustainable packaging, of course).

LIMITING THE PROMOTION AND MARKETING OF UNHEALTHY FOODS

Techniques to market and advertise foods can influence consumer behavior in positive or negative ways. Examples include social media, print and television advertising, in-school marketing, toys and products with brand logos, packaging, and product placements. Television ads are particularly influential, as advertisers often use child-oriented persuasion to promote junk food, that inevitably makes kids beg their parents to buy these unhealthy processed foods. Governments can intervene in schools and protect children by banning food industry companies from sponsoring sports programming or supplying vending machines stocked with their products.

Advertising campaigns for unhealthy foods, especially those marketed toward children, should be eliminated. Mothers should be protected from aggressive infant formula marketing practices that try to urge them to use their products instead of breastfeeding, going against global recommendations of the World Health Organization. This can be accomplished through large-scale education campaigns, by excluding the formula industry from nutrition education and policy roles, and by imposing strong penalties for violations of the International Code of Marketing of Breast-milk Substitutes. Across the board, junk food advertising and other forms of commercial promotion should be restricted when they target children and vulnerable populations or undermine public health policy.

Food and beverage companies see marketing and advertising, product placements, pricing policies, and packaging as a response to consumer demand. This view puts the responsibility solely on the consumer to make the "right" choice, even though the present balance of power highly favors multinational food and beverage corporations. Companies spend a fortune on behavioral studies and focus groups that inform how they can most effectively sway or manipulate shoppers through advertising and packaging decisions. In addition, these businesses argue that processed foods are required to feed a growing, urbanizing population, many of whom have rising incomes and are demanding greater convenience. Efforts to create healthier food environments for consumers should redress the power imbalance between consumers and industry to give consumers more agency and awareness in their food choices. Governments need to step in.

Currently, some of the Scandinavian countries, the United Kingdom, the Netherlands, and other countries have advertising laws covering food. The United Kingdom restricts advertising of junk food to children; the Netherlands bans all forms of food advertising for children under age 13; and France requires warnings on advertisements for unhealthy foods. In 2006, Brazil attempted to enact anti–junk food advertising laws, along with other measures to curb obesity and disease, but failed because of industry opposition. In the United States, the food industry self-regulates advertising to children, which is largely ineffective in protecting consumers.

CONSIDERING THE EAT-*LANCET* REPORT RECOMMENDATIONS

In an effort to increase public awareness of the challenges facing food systems, the *Lancet* journal published the EAT-*Lancet* Commission Report on Food, Planet, and Health, on which I was privileged to serve. The commission brought together 19 commissioners and 18 coauthors from 16 countries in various fields of human health, agriculture, political science, and environmental sustainability. The commission was chaired by two heavyweights in their respective fields—Johan Rockström, a specialist on global sustainability, and Walter Willett, a nutritional epidemiologist at Harvard University. Over the course of two years, we worked to determine whether a diet is possible that can maintain and improve human health while remaining within boundaries of planetary sustainability. The result was "Our Food in the Anthropocene: The EAT–*Lancet* Commission on Healthy Diets from Sustainable Food Systems," published in 2019.[16]

This report was notable in many ways. It was the first scientific review of how to achieve a healthy diet from a sustainable food system. It set scientific targets, forged consensus, inspired organizations, got people thinking about nutrition, and sparked scientific and political debate. The report called for a "Great Food Transformation," stating that "delaying action will only increase the likelihood of serious, even disastrous consequences."[17] It concluded that global consumption of fruits, vegetables, nuts, and legumes will have to double and that consumption of foods such as red meat and sugar will have to be reduced by more than half from current levels.

Most important, the report described a universal health reference diet aimed at meeting the nutritional needs of the planet's future population while limiting global temperature increase to 1.5 degrees Celsius, as specified in the Paris Climate Agreement. Also called the Planetary Health Diet, this eating plan calls for an increased consumption of healthy foods such as fish, vegetables, fruit, legumes, whole grains, and nuts and a decreased consumption of unhealthy foods such as red meat, sugar, and refined grains. Optional foods to be consumed in moderation include eggs, poultry, and dairy foods. The diet advocates for reasonable caloric intake, with consumption not to exceed 2,500 calories per day. Here is a graphic representation of the Planetary Health Diet.

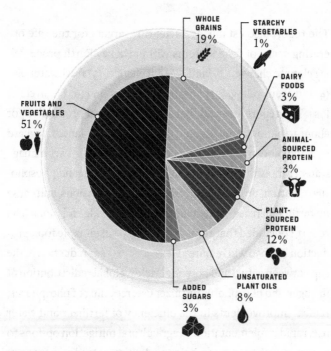

WHOLE GRAINS
19%

STARCHY VEGETABLES
1%

DAIRY FOODS
3%

FRUITS AND VEGETABLES
51%

ANIMAL-SOURCED PROTEIN
3%

PLANT-SOURCED PROTEIN
12%

ADDED SUGARS
3%

UNSATURATED PLANT OILS
8%

EATING FOR PERSONAL AND PLANETARY HEALTH: The Planetary Health Diet recommended by the EAT-*Lancet* Commission on Food, Planet, and Health consists of approximately half a plate of vegetables and fruits; the other half should consist of primarily whole grains, plant protein sources like legumes (beans, lentils, peas) and nuts, unsaturated plant oils like olive oil, and, optionally, low to modest amounts of high-quality animal-source foods. Added sugars and starchy staples like corn, potatoes, and rice should be minimized. Despite a number of shortcomings, this diet laid out in the EAT-*Lancet* Commission report was highly influential and grounded in what most national dietary guidelines suggest.

W. Willett et al., "Our Food in the Anthropocene: The EAT-*Lancet* Commission on Healthy Diets from Sustainable Food Systems," *The Lancet* (2019): 1-47.

The report also established scientific targets for the safe operating space of food systems within six key Earth processes: (1) climate change, (2) land-system change, (3) freshwater use, (4) nitrogen cycling, (5) phosphorus cycling, and (6) biodiversity loss. The report asserted that sustainable food production for about 10 billion people should use no additional land, safeguard existing biodiversity, reduce consumptive water use, manage water responsibly, substantially reduce nitrogen and phosphorus pollution, produce zero carbon dioxide emissions, and cause no further increase in methane and nitrous oxide emissions. The report concluded that transformation to sustainable food production by 2050 will require at least a 75 percent reduction in the gap between actual and potential yields, global redistribution of nitrogen and phosphorus fertilizer use, recycling of phosphorus, radical improvements in the efficiency of fertilizer and water use, rapid implementation of agricultural mitigation options to reduce greenhouse gas emissions, adoption of land management practices that shift agriculture from a carbon source to a sink, and a fundamental shift in production priorities.

The EAT-*Lancet* report filled a critical gap in global nutrition and environmental policy. However, it also received significant criticism when it was released. I participated in one of many global launch parties in Rome, where I sat next to the president of the International Fund for Agricultural Development, Gilbert F. Houngbo. As a citizen of Togo, he told me that the report did not accord with the situation facing African countries because of its high-income view of the world. He's not entirely wrong. Other opponents of the report attacked the Planetary Health Diet, calling it unaffordable, protein deficient, and insufficiently supported by science. A study published after the Commission report found that nearly 1.6 billion people don't have the financial means to follow the Planetary Health Diet, which is especially concerning given that malnutrition is concentrated among economically poorer populations.[18] Other critics cited practical challenges that may hinder implementation of the Planetary Health Diet and argued that this diet does not account for cultural differences in diets around the world. Because the diet recommended specific ranges of food groups, many felt it was overly prescriptive. Furthermore, meeting the diet's nutritional needs would require doubling yields of fruits, vegetables, and nuts globally by 2050, but the effects of climate change on agriculture will make it difficult to achieve this goal.

The report was especially contentious because of its potential effects on the livestock sector. It suggested a significant scaling back of livestock production systems. Opponents acknowledged that significant reductions in meat consumption, as the report advised, could benefit populations in high- and middle-income countries. However, they pointed out that animal-source foods can be a valuable source of nutrients for people experiencing undernutrition. And of course, those working in the meat industry are critical of the report as it directly affects their livelihoods.

Other criticisms of the report were that it didn't set a time frame for when the suggested targets should be reached or a price tag on how much reaching those targets would cost. Nor did it take on the entirety of food systems (just the two ends of food systems—agriculture production and consumption), thoroughly examine the behavior of stakeholders, or probe the economic impacts of its suggested changes to food systems.

Despite its shortcomings, the report spurred countries to take a much deeper look at their food systems in the context of climate change. It prodded them to answer these questions: If your country were to have a sustainable diets policy, what would that look like? What would that mean for your agriculture and food production sectors? How would diets need to change? And who would benefit? Above all, the report asked each country two major questions: (1) What needs to change in your country to create a sustainable food system? (2) What trade-offs are you willing to live with to save yourselves and the planet?

MOVING TOWARD HEALTHY, SUSTAINABLE CHOICES

One way to improve dietary choices is to re-evaluate the label information about a food product and any declarations on its packaging and provide national food-based dietary guidelines. Nutrition labels are effective for both food producers and consumers, as they encourage healthier individual choices and prompt the food industry to reformulate products with nutritious ingredients. This information is often found on the back of the package. However, many products carry misleading front-of-package claims on the health and/or nutrition benefits of foods — an unregulated area of packages. Producers often design the product's packaging to ensure that these statements are immediately seen by the consumer, who then may or may not flip the package over to evaluate the fine print. These marketing strategies contribute to unsustainable food environments in which consumers think they know what they're eating but are continually deceived.

The Codex Alimentarius Commission established by the United Nations has developed standards for nutrition guidelines on food products. However, these labels require some degree of nutritional literacy, need to be culturally understood, and are overwhelming or difficult to interpret for many people. For this reason, there have been recent moves to adopt easy-to-read and interpret labels (e.g., traffic lights, star ratings) on the front of the package or on store shelves. Consumers can easily interpret graphic front-of-package labels that incorporate colors, symbols, and text to indicate nutrition or health compared with labels that only emphasize numeric information, such as daily recommended amounts expressed as grams or percentages. Labels of this type (as long as they're not deceptive) would be easier for consumers to interpret and could lead to better food choices.

In 2012, Chile became one of the first countries to implement front-of-package labels with black-and-white stop signs placed on foods with excessive fat, sugar, salt, or calories. In addition, new requirements prevented those foods from being sold in or near schools or advertised during prime-time television. Since its implementation, 93 percent of the population of the metropolitan region of Santiago de Chile has reported recognizing the front-of-pack nutritional warning, and 92 percent say that these warnings have influenced their purchasing decisions.[19] Overall, Chile has had a 23 percent decline in the total purchase of foods with front-of-package warnings.[20] These bold actions have inspired other countries to adopt similar labeling. Dr. Ricardo Uauy, a leading proponent of these measures who advocated for years to see them implemented, described the work this way:

> The progressive adoption of this nutritional information on packaging was sparked by the crisis of rapidly rising obesity rates in our children and adolescents. We worked with influential entities including the Health Committee of the Senate and the Ministry of Health to limit the food industry's inclusions of saturated fats and added sugars and sodium, and invited the industrial/business sector (food companies, retailers and advertisers) to participate in the process—but not to control the changes we intended to advance. Some of these partners resisted the changes, rejecting the limits and critical nutritional

levels we defined, but the warning labels were ultimately instituted, and the Health Committee of the Senate continues to guide legislative changes supporting calorie reduction.[21]

Governments and businesses can also work together to provide messaging on the importance of nutrition and the benefits of certain foods. In the United States, food-labeling legislation has had some success in encouraging the food industry to develop healthier products rather than face the stigma associated with certain labels. This has best been demonstrated with trans fat labeling laws, which went into effect in 2006 and resulted in a significant reduction in trans fats in the food supply, which contributed to a decrease in cardiovascular disease.

Food-based dietary guidelines can steer people toward healthier and sustainable food choices by increasing consumer knowledge and awareness. First, guidelines can provide a unified voice to the public on where the government stands on the latest dietary advice in the context of health promotion and disease prevention. Second, they serve as the foundation for food and nutrition policies instituted within a country and guide budgetary allocations for such programs as school lunches. Third, the food and beverage industries often respond to changes proposed in dietary guidelines by reformulating products and answering to consumer demands. Many recommendations extend across countries, such as suggestions to consume a variety of foods; to consume fruits and vegetables, legumes, and animal-source foods; and to limit salt, sugar, and fat. However, few guidelines address environmental factors such as greenhouse gas emissions and water pollution, or sociocultural factors such as labor conditions.

Unfortunately, only a handful of countries have guidelines that specifically promote both healthy and sustainable diets and food systems.[22] A recent study examined 83 food-based national dietary guidelines and found that 98 percent were not compatible with at least one of the World Health Assembly Action Agenda on Non-Communicable Disease targets, the Paris Climate Agreement, or Aichi Biodiversity targets.[23] In some cases, governments actively oppose the inclusion of environmental sustainability into dietary guidelines. In a press statement responding to the 2015 US Dietary Guidelines, the US Secretaries of Agriculture and Health and Human Services stated: "We do not believe that the 2015 DGAs (Dietary Guidelines for Americans) are the appropriate vehicle for this important policy conversation about sustainability" and that the purpose of dietary guidelines was simply to educate the population about weight control and chronic disease prevention, not sustainability. The 2020 guidelines do not include any sustainability aspects to their guidance. In contrast, Germany's dietary guidelines urge consumers to "choose mainly plant-based foods. They have a health promoting effect and foster a sustainable diet."[24]

I contributed to a 2018 research study examining global dietary guidelines to see how much they integrated human health and environmental sustainability.[25] Australia, Brazil, Canada, Qatar, and Sweden scored high in the degrees of sustainability. Brazil's guidelines, for example, focus on meals and encourage citizens to simply cook whole foods at home and to be critical of the manipulative marketing practices of big food corporations. While America's dietary guidelines divide foods into "good" foods and "bad" foods, with a de-humanizing emphasis on nutrients and food groups over meals, Brazil's guidelines emphasize the human side of food consumption—eating as a social experience. The guidelines culminate in Brazil's "golden rule," which states: "Always prefer natural or minimally processed foods and freshly made dishes and meals to ultra-processed foods."[26] Brazil's rules were revolutionary in that they framed unhealthy foods less in terms of their nutritional composition and more in terms of the degree to which they have been processed. Considering the world's increasing consumption of highly processed foods and their resulting health problems, this is a major step toward healthier diets.

After the success of Brazil's new dietary guidelines, Canada similarly revamped its food guide in 2019. The new recommendations are visually represented by a plate half-filled with fruits and vegetables, a quarter-filled with protein foods, and a quarter-filled with whole grain foods, with water as the drink of choice (much like the EAT-*Lancet* Planetary Health Diet). Like Brazil's dietary guidelines, the document focuses on the social aspects of eating, reminding Canadians to cook more often, eat meals with others, be mindful of their eating habits, and enjoy food. It also advises consumers to read nutrition labels, be aware of food marketing, and limit foods high in sodium, sugars, or fat. These new guidelines are a departure from Canada's prior recommendations, which included more animal-source foods and refined grain products and were heavily influenced by the food industry.

STRENGTHENING FISCAL POLICIES

Policy makers need to create strong fiscal frameworks that shape the actions of those responsible for our food systems. Tax, subsidy, and trade policies all need to better align with policies that promote healthy and sustainable diets. Industry goodwill and voluntary measures will not be enough. While some in the food and beverage industry are acting in ways that benefit public health, transgressions against public health goals remain common. Furthermore, only governments have the necessary legitimacy to establish a fiscal framework that puts diets on a healthier and more sustainable track.

Agriculture subsidy policies are not aligned to what constitutes a healthy diet. Most subsidy policies focus on the major staple crops such as corn, soy, rice, and cotton, along with input subsidies such as fertilizer (which has been adopted in many African countries like Malawi and Tanzania). Aligning agriculture subsidy programs toward commodities that support healthy diets, such as fruits, vegetables, nuts, and legumes could be a game changer for farmers who rely on those subsidies year in and year out, and for consumers to see shifts in what is prioritized across agriculture systems where they live.

Local or national governments could institute tax incentives to motivate producers and retailers to engage in healthier and more sustainable practices. Governments could tax fertilizer, which could encourage farmers to switch to more organic approaches. Governments could also use tax funds to pay premiums to wet market retailers if they meet minimum food safety standards. They could provide incentives to street vendors to use healthier ingredients in exchange for discounted ingredients and certifications as they did in Singapore with street food hawkers; give tax breaks or financial incentives for store retailers to sell healthy foods; or incorporate tax rates that incentivize more nutritious food products. Efforts to encourage corner stores to stock healthy, fresh foods have increased purchases of these foods along with higher profits. New York City's Healthy Bodega program has linked bodegas selling healthy foods for consumers to the social safety net program, SNAP. The local production and sale of healthy foods and direct sales through farmers markets and Community Supported Agriculture offers important economic and social benefits to farmers, consumers, and communities, particularly in neglected and impoverished places.

One valuable way to influence consumers' diets is to make unhealthy foods more expensive and nutritious foods cheaper.

Studies have shown that sugar-sweetened beverage purchases can be reduced around 10 percent through taxation, while subsidizing vegetables and fruits can increase consumption by 10 to 30 percent.[27] Imposing substantial taxes on fattening foods could improve health outcomes, but this remains difficult to implement.

However, taxes and subsidies have the potential to further inequities by imposing a larger burden on the poor than the rich, who already pay much more for groceries proportionate to their incomes than others with more means. About 60 countries have already imposed taxes on sugar-sweetened beverages. These soda taxes can be considered "regressive" in economic terms because poor people tend to consume more of these beverages. Although the soda tax has been effective in reducing consumption, policy makers also need to develop progressive taxes in which wealthier people pay the larger share if there are taxes on foods that may have some benefit, such as a carbon tax on red meat that could go toward city or national climate change adaptation and mitigation strategies. A potential solution to this problem is to add food subsidies to taxes or a higher tax on expensive luxury items purchased mainly by the most wealthy to alleviate potential regressivity and enable consumers to switch to healthy foods without incurring additional costs.

In 2013, Mexico instituted an 8 percent tax on all "nonessential" foods, including snacks, sweets, nut butters, and cereal-based prepared products. Within these categories, foods that exceed a threshold of more than 275 calories per 100 grams are taxed. Studies have shown that these policies have changed people's eating habits for the better. One study of the Mexico junk food tax found that people purchased 7 percent less sugar-sweetened beverages than they would have without the tax.[28]

Tax policies can also support greater environmental health while making diets more sustainable and nutritious. A greenhouse gas emissions tax on foods corresponding to their emissions intensities could be a powerful health-promoting climate policy, and the tax income could be used to subsidize healthier foods or go back into the health care system. Other tax policies related to health and sustainability include a water use tax, a meat tax, a carbon tax, a pollution tax, and a sugar tax. One type of carbon dividend proposal aims to tax carbon when it comes out of the ground and then to equally distribute the returns among all people.

Finally, governments can influence health and sustainability outcomes through their regulations and policies related to international trade. A few years ago, my colleagues Steve Wood at Yale University, Ruth DeFries at Columbia University, and I published a study that showed that international trade enables the global food supply to better distribute nutrients around the world.[29] Without trade, 934 million people across the world could be deprived of protein and at least 146 million would not be able to fulfill their vitamin A requirements. Trade can expand the variety and distribution of foods around the world, lower food prices, and extend the number of days per year that products are available.

Trade can also create large flows of unhealthy foods. Governments must consider the relationship between trade and nutrition to determine the best policies given their context. If they decide to create policies that open their markets and integrate them into global trade, governments must be especially responsive to the needs of poor consumers and resource-scarce producers. On its own, the market will not protect vulnerable consumers who do not have sufficient purchasing power to achieve a healthy diet. The far-reaching supply chains of multinational companies like PepsiCo could be leveraged and incentivized to not only deliver soda, but healthy foods and health commodities (like vaccines).

STRENGTHENING THE EVIDENCE AND DATA TO INFORM DECISION MAKERS

Many nations that lack good nutrition and sustainability policies blame a lack of evidence to support legislation. However, action generates evidence, and a lack of evidence is no excuse for inaction. We have no time to waste; governments need to act now.

At the same time, governments need to invest in research and development on food systems to identify promising and proven policies and programs that could be scaled up to support healthy and sustainable diets. A particular need is for metrics and data on both sustainability and health. Indicators should range broadly, including long-term ecosystem health, total resource flows, interactions between agriculture and the wider economy, the sustainability of outputs, nutrition and health outcomes, livelihood resilience, and the economic viability of farms.

Dietary data also needs to be collected across all countries, with more disaggregation of socioeconomic status that takes equity issues into account. Most dietary data today comes from high-income countries. Focusing on low- and middle-income countries is critical, since these are the populations who are most vulnerable to malnutrition, hunger, and the effects of climate change. There's a renewed focus now to collect dietary data around the world. Significant dietary projects such as Tufts University's Global Dietary Database funded by the Bill & Melinda Gates Foundation and the FAO's GIFT program will be "beefing" up dietary data that is inclusive of low-, middle-, and high-income countries in a streamlined way.

One recent solution to this dearth of data is an easy-to-navigate online tool—the Food Systems Dashboard—developed by my team at Johns Hopkins University, the FAO, and the Global Alliance for Improved Nutrition (GAIN). It's designed to help decision makers understand the food systems, identify the levers of change, and decide which ones to pull. The dashboard is a unique, holistic resource intended for policy makers, nongovernmental organizations, businesses, civil society leaders, and other actors to enable timely visualization of national food system data, to understand the interconnections across multiple sectors, to perform comparisons with other countries, identify key challenges, and prioritize actions.

What struck us back in 2017 while working on the *UN High Level Panel of Experts on Food Systems and Nutrition Report* was the lack of accessible, organized, quality-checked information on food systems. Without that data, it's difficult to identify the best evidence-based actions that could improve food systems. It was really important to us, given the level of complexity and interconnections inherent to food systems, to present data in a way that's easily digestible, and that's what the dashboard does. Now decision makers have easy access to both data and to policy advice that is specific to their situations.

The dashboard, launched in June 2020, houses open-source food systems data of more than 230 countries and territories by bringing together more than 170 indicators from 35 sources. It enables stakeholders to compare their food systems with those of other countries and provides guidance on potential priority actions to improve food systems' impacts on diets and nutrition.

My colleague Lawrence Haddad, the executive director of GAIN and winner of the World Food Prize in 2018, has said: "The dashboard has the potential to halve the time required to gather the relevant data, helping public agencies and private entities to grasp the three Ds more rapidly: Describe national food systems, Diagnose them to prioritize areas for action, and then Decide on the action to take based on plausible interventions that have been tried in other countries."[30]

How data is communicated and used is also important in changing public health policy. Many years of strong science have shown that trans fats are unhealthy and lead to many negative health outcomes. When New York City Mayor Michael Bloomberg learned of evidence that trans fats increase rates of cardiovascular disease, morbidity, and mortality, he decided to enact a ban on trans fats in restaurants in New York City in 2006. His actions drew national attention, and other states started to follow suit. Then, in 2018, the US Food and Drug Administration announced that companies would have one year to replace trans fats in their food products with other ingredients, effectively banning trans fats from the entire American food system.

The elimination of trans fats in America is a great example of how science can inform local policy, which in turn can inform national policy. The private sector was willing to transform food production practices, though they were aided by the fact that relatively easy replacements (unfortunately, often palm oil, which is responsible for significant deforestation and air pollution in Southeast Asia, is not the healthiest oil) are available in lieu of trans fats. Nevertheless, the success of the trans fats ban reveals that clear evidence can result in food policy and regulatory changes at the highest level of government.

Smoking is another case in which clear evidence produced striking governmental action. After decades of evidence demonstrated the deleterious effects of smoking, the US government enacted legislation to place large, graphic warnings on the front of cigarette packages. Instead of reading, "smoking *may* be a risk for mortality," they simply say, "smoking kills," because the evidence is so clear. Packages now include close-up images of teeth, lungs, and throats that have been destroyed by cigarette smoking. Governments also tax cigarettes, restrict their advertising, and prohibit where you can smoke to protect public health. The nutrition sector needs concrete evidence like the evidence found for tobacco use to convince policy makers of the seriousness of these issues and to take action. Good data can produce better awareness of what has worked in nutrition and can result in actionable policies.

Finally, governments must support sharing data and knowledge. In partnership with researchers, policy makers need to foster standards for data collection and sharing in public institutions and other settings that support health and sustainability throughout the food chain. At the same time, policy makers and researchers need to work with the system, rather than against it, by using what's already there. Accomplishing these goals will require better diagnostics and surveillance, expanded delivery platforms, and stronger, open-source data systems.

Investing in metrics and data to create informed policies will help improve health and nutrition outcomes. But governments must also work with the data that already exists to create immediate, evidence-based strategies aimed at creating more equitable, sustainable, and healthy food systems. Future data will help adjust and improve government policies around nutrition, but the time to act is now.

The Food Systems Dashboard: A New Global Information Tool

Launched in June 2020, the Food Systems Dashboard provides a comprehensive, convenient, and interactive way for 230 nations to evaluate their food systems from a myriad of perspectives, consider specific policy advice to clarify forecasting, and prioritize actions. It's "one-stop shopping" for evidence-based food systems assessment and decision making that drew 50,000 people to the site in the inaugural months. Here are a few key features:

- Brings together diverse data from more than 170 indicators (what people are eating, greenhouse gas emissions coming from agriculture, nutritional deficiencies and disease concerns, etc.) for every country and territory, from 35 public and private sources including UN agencies, the World Bank, the Food and Agriculture Organization, and Euromonitor.
- Policy makers, nongovernmental organizations, and private companies can cut their data-gathering time in half, see previously undiscovered interconnections of factors across food systems, and have access to solutions that other nations have piloted, fostering collaborations around the world.
- Information is presented visually with icons and easy to read graphics to aid quick engagement with the information.
- Access to the website is free of charge to everyone: farmers, food producers, transport companies, and policy makers.
- While introduced in English, the data will be presented in French and Spanish in 2021 (and additional languages, as funding permits).

FINANCING FOR NUTRITION
AND FOOD SYSTEMS

———

Beyond specific policies, the overall amount of money being devoted to improving nutrition is far too low. At present, only 2 percent on average of general government expenditures worldwide are spent on interventions to address undernutrition.[31] Funding to reduce obesity and diet-related noncommunicable diseases is also insufficient, receiving less than 2 percent of development assistance for health.[32]

If governments were to commit much more financing to improving diet-related health outcomes, reduced rates of malnutrition would produce substantial returns and decrease health care costs. Devoting just US$7 billion annually to nutrition could globally reduce the number of stunted children by 40 percent, reduce the number of women of reproductive age with anemia by 50 percent, increase the rate of exclusive breastfeeding by up to 50 percent, and reduce child wasting to a level of less than 5 percent.[33] Research has shown that investments in nutrition provide an estimated return between US$4 and $35 for each US$1 invested worldwide.[34]

The policy agenda is ambitious. We need to take a hard look at how to fully integrate sustainability and health issues into food policies. At the same time, policy makers must increase their understanding of the issues involved in nutrition and sustainability. Nutrition experts need to better understand the policy process that exists and the way that it works so that they can get involved, offer their expert advice, and leverage the greatest impact.

No one measure can successfully drive these necessary changes. A constellation of different approaches and strategies, scaled and operating across supply chains targeted to the full spectrum of people and organizations are needed. Governments, industry, and citizens need to care about diets, nutrition, climate change, and food systems. We then need to act to translate these goals into reality.

CONSIDERING TRADE-OFFS

Governments will need to act quickly, partner well, stay informed, and be efficient in creating strategies and policies. They also should consider dangers of trade-offs when formulating food policies. Palm oil is an example of a food that has trade-offs. Palm oil can serve as a replacement for trans fats, which are known to harm human health. But at the same time, as mentioned, palm oil leads to increased deforestation and loss of biodiversity—specifically, to orangutan habitats.[35]

My former postdoc, Shauna Downs, now a professor at Rutgers University, examines the effects of the palm oil industry in Myanmar. She found that

in the early 2000s, the government of Myanmar began providing incentives for the domestic production of palm oil. Through my work examining the trade-offs related to palm oil production and consumption in the country, I've found that its production has limited economic viability, it contributes to environmental degradation, and threatens the land tenure of socially disadvantaged groups. From a health perspective, it's high in saturated fat and its use is ubiquitous in highly processed foods. While consumers don't want to consume it because of concerns related to its quality, its low cost and widespread availability leads to its consumption, particularly among lower-income groups. This work has highlighted the importance of applying a more holistic approach to assessing the policies that drive what we eat. We need to go beyond looking at these policies from an economic perspective and consider the real health, sustainability and social trade-offs of their adoption.[36]

Even the Mediterranean diet, which is often considered heart healthy and sustainable, has environmental trade-offs. It promotes olive oil and nuts as a source of healthy fat, but olive and nut trees require large amounts of water to grow.[37] Similarly, almond milk, a substitute for lactose-intolerant people, also requires an enormous use of water to produce. Such environmental, health, and economic considerations must be weighed to determine the best policies to balance planetary benefits and human well-being.

Many structural interventions implemented by governments have nudged food systems and environments toward better nutrition and environmental sustainability. In Britain, schools must provide fruits and vegetables, and high-quality meats and cereals. No potato chips, chocolates, sweets, or drinks with added sugar are allowed in school meals or vending machines, and fried food is limited to twice per week. As discussed earlier, Chile has mandated warning labels on the front packages of unhealthy foods and Norway has regulated junk food marketing and advertising to children. South Korea has implemented fast food–free zones around schools. In India, governments are supporting more sustainable, organic production practices on farms. There are so many examples out there but still, it's not enough. There needs to be stronger political will, more investment, increased inclusiveness, and better governance and accountability of food system actors.

Over the past decades, the concentration of economic power in private companies has restricted the political power of local and national governments to take action in supporting healthy and sustainable diets. Moreover, the rising influence of multinational corporations has stymied efforts by public health officials to create healthier food environments. Truly changing nutrition and environmental outcomes will require commitments by both the government and the food and beverage industries. Obtaining this commitment from those industries will be a major challenge, but it's of the utmost importance in moving toward better food systems — and a better protected planet.

CHAPTER 5

Can One Bee Save the Hive?

WE ALL MAKE DECISIONS ABOUT FOOD on a daily basis, but the choices we make are not ours alone. Our decisions involve a complex interplay of images, memories, and emotions that inform us subconsciously and subliminally. We each have our daily pressures and stresses that leave little time or mental space for exercise or preparing and cooking a hot meal. We're influenced by the settings in which we make choices. We're surrounded by salty and sugary snacks and are often unaware of the undesirable health and environmental impacts connected to what we buy, even seemingly healthy items. Many of us live in food environments where it's hard to make the right choices, and many people have far fewer choices and available resources than others.

Given all these factors, food choices can seem overwhelming, monotonous, or nonexistent, depending on the context. Health and sustainability are not and should not be the sole burden of the individual. What we eat and what policies we support shape food systems and the food supply. Individual actions can contribute to and support much larger social movements that collectively shift the food agenda through the media headlines they prompt.

Transitioning to healthy and sustainable diets isn't easy. It requires knowledge, will, and persistence. What works for one person may not appeal to or work for another for a host of reasons. But no matter who we are, we all should have individual opportunities to support better health and sustainability for ourselves, our families, our communities, and the planet.

MOVING TOWARD HEALTHY, SUSTAINABLE DIETS

There is no single quality diet, because diets depend on individual needs and physiology, culture and social norms, local food availability and accessibility, and dietary customs. Nevertheless, there is a general consensus on what constitutes a healthy diet:

- A sufficient quantity and balance of macronutrients and micronutrients to maintain life, support physical activity, and achieve and maintain a healthy body weight.
- A diversity of nutrient-dense foods, such as vegetables, fruits, whole grains and cereals, dairy, and animal- and plant-based proteins, appropriate to a given geographical location and cultural context.
- A balance of foods, with only moderate amounts of processed animal-source foods and with limited

consumption of nutrient-poor foods associated with
adverse health outcomes, including obesity and its
associated noncommunicable diseases.

- Foods that are safe, without contamination from harmful
bacteria, viruses, parasites, or chemical substances during
production, storage, distribution, and preparation.

Eating a quality diet has the potential to not only improve human health but also simultaneously protect the environment. In particular, three specific measures could go a long way toward improving both human and environmental health.[1]

First, we need to reduce the overconsumption of calories; for adult women, that's 1,600 to 2,400, and 2,000 to 3,000 for adult men. Moderate consumption entails eating to satisfy but not exceeding energy and nutrient requirements for growth, activity, and bodily repair. Moderate consumption typically results in achieving and maintaining a healthy weight, thus avoiding health risks associated with obesity. In addition, consuming no more than our nutritional requirements places less demand on finite resources to produce, process, and distribute extra food.

Second, we need to avoid unhealthy, highly processed foods. This can be complicated, especially in high-income, industrialized countries, where nearly everything has been processed to some extent. However, the foods to avoid are those with ingredients rarely or never used in kitchens. In addition to reviewing food labels, consumers can check out the website Open Food

Facts, which classifies foods by the NOVA method, a four-category system that identifies the degrees to which foods have been processed.

We often blame ourselves for overindulging in these foods, but larger factors are at play. In addition to the marketing and advertising aimed toward getting us to buy such foods, highly processed foods containing high amounts of sodium, sugar, and unhealthy fats are associated with better taste, pleasure, craving, and loss of control. In addition, modern patterns of work make it difficult to find the time to cook and eat a high-quality diet. For many of us, entirely eliminating processed foods is impossible because of their ease and their affordability. However, reducing consumption of those highly processed foods that provide very little health benefit could significantly improve health outcomes while supporting environmental health.

Third, individuals in middle- and high-income countries need to reduce their consumption of animal-source foods, beef in particular. Even as the majority of people in high-income countries consume far more meat than they need, most people in low-income countries don't consume enough animal-source foods. These foods need to be more accessible and affordable to people in low-income countries so that everyone has the opportunity to get the nutritional benefits that animal-source foods can provide. In low-income countries, the aim should be to consume enough meat to fulfill nutritional needs. That means improving food supply chain infrastructure and subsidizing prices to ensure these foods are affordable. In middle-income countries, the aim should be to prevent meat consumption levels from reaching excessive levels.

BALANCING SUSTAINABILITY AND HEALTH IN FOOD CHOICES

As individuals, we owe it to ourselves to eat a healthy, high-quality diet. As earlier chapters highlighted, this can be difficult for many people around the world because of systemic structural inequities and a lack of governance of food systems. At the same time, eating a healthy, high-quality diet is not always straightforward, particularly if we incorporate environmental sustainability choices into the mix.

Diets with high amounts of dairy, lean meat, fish and seafood, nuts, and fruits and vegetables may meet individuals' nutritional needs, but some of these foods can have a high environmental footprint, as the example about bananas in the introduction attests. In addition, certain fish and seafood production practices, such as trawling, contribute to greenhouse gas emissions, reduce marine biodiversity, and destroy the ocean floor, not to mention that they can lead to unsafe or unjust labor conditions. Cultivation of some nuts including almonds and cashews have a high water use footprint, whereas walnuts and sunflower oil have lower footprints, and fish and other seafood can be produced sustainably through responsible aquaculture. But, as individuals, making sustainable choices requires informing ourselves about sustainable practices and applying this knowledge to our food choices. This can be incredibly time-consuming and challenging at first given all the mixed, confusing messages out there, and the sheer volume of products we have to choose from in our stores.

For consumers to make knowledgeable food choices, the research community and educators must effectively communicate the science behind those choices. It requires governments to take up the evidence and make policy decisions that provide consumers with more well-informed and equitable choices. It requires that those in the food industry supply information about their production practices to the public, whether on labels or online. It requires that media outlets publish simple, easy to digest messages and ensure that they get the science right. Without such guidance, it's difficult to know what is or is not sustainable or healthy.

Campaigns and mass media can help in raising awareness for consumers. National school meal programs can also promote existing campaigns like Meatless Mondays to nudge kids toward more vegetarian options. In high-income countries, the aim should be to alter entrenched patterns of excessive consumption. This could involve taxing high-carbon foods, encouraging fast-food chains and restaurants to include more alternative protein options, and requiring public procurement facilities to include more plant-based meals in schools, hospitals, and prisons.

When health and sustainability align, choices become easier. A lot of research has been done to assess the impacts of specific foods and food groups on human and planetary health. Fish is generally a healthy choice but has a bigger environmental footprint on average than plant-based diets, and larger ocean species tend to have high levels of mercury. Producing unprocessed red meat has the highest impact for most environmental indicators. Foods with medium environmental impacts or not significantly associated with ill health—such as dairy, eggs, and chicken—could help improve health and reduce environmental harm if they replaced foods such as red meat.[2] Shifting toward consumption of milk and yogurt would allow consumers to retain the health benefits of dairy foods while reducing their environmental footprint.[3]

Kenyan food systems illustrate the many factors that influence attempts to achieve healthy and sustainable diets. Take camel milk—today its production is primarily a low-tech business, yielding an average of 5 liters per animal per day. Improved breeding and husbandry could raise production to 20 liters per day. This could help reduce the burden of malnutrition in the Horn of Africa, although doing so will require overcoming the challenges posed by recurrent harsh droughts and possibly cultural preferences.

Rather than following a prescriptive plan, regions must consider how to adapt their current food systems to promote greater human and environmental health. As individuals, we must work within the food systems in which we live to make the best choices possible. The foods available in Kenya are very different than foods available in the United States. Nonetheless, for most people, choices exist in both places, some of which are healthier and more sustainable than others. Ensuring these foods are appealing, tasty, and at the right price point has to be prioritized as well.

Reining in climate change will not require completely eliminating foods with high environmental impacts. Merely curbing consumption and replacing some high-impact foods with low-impact substitutes could result in significant benefits. The World Resources Institute argues that if beef consumption in high-consuming countries declined to about 50 calories a day it would nearly eliminate the need for additional agricultural expansion and associated deforestation.[4] The institute also reports that Americans could nearly halve their diet-related environmental impacts just by eating less meat and dairy.[5] Plant-based burgers, finless fish (seafood grown from stem cells), and blended meat-plant alternatives are also beginning to make an appearance at grocery stores and fast-food restaurants. These products mimic meat in their taste, texture, and look.

Globally, transitioning to diets that emphasize plant-based foods could reduce global mortality by 6 to 10 percent and food-related greenhouse gas emissions by 29 to 70 percent compared with the current trajectory scenario up to 2050.[6] More than anything, individuals (particularly those who live in high-income contexts who consume high amounts of meat) can help support the planet and their own health by focusing on plant-based foods and consuming low to moderate amounts of animal-source foods.

CHOOSING DIETARY PATTERNS TO FIT LIFESTYLES

People often ask which well-known diets are optimal for health and sustainability. Typical dietary patterns include the vegetarian diet, the vegan diet, the flexitarian diet, the pescatarian diet, and the Mediterranean diet.

Because replacing animal-based foods with plant-based alternatives confers the greatest environmental benefits, vegan and vegetarian diets are associated with the greatest relative reductions in greenhouse gas emissions, water use, and land use.[7] Diets that have more fish and poultry and less meat from ruminants also result in decreased environmental impacts but less so than vegetarian diets.

To consume balanced levels of protein, vegetarians should include a variety of protein-rich plant foods such as nuts, seeds, legumes, tempeh, tofu, and seitan. In addition, they have the option to include eggs and dairy-based products to meet their protein and nutrient needs. The vegan diet is similar to vegetarianism, but animal-source foods are entirely eliminated. For vegans, it's especially important to consume a variety of plant-based foods and protein-rich plants to meet nutritional requirements.

The flexitarian diet advocates for consumption of primarily plant-based foods, though meat and other animal products are allowed in moderation. The pescatarian diet is similar to the flexitarian diet, but it only allows for seafood consumption and no other meats. Some pescatarians also eat dairy and eggs (in which case they are technically lacto-ovo-pescatarians).

Finally, the Mediterranean diet emphasizes plant-based foods such as fruits, vegetables, and legumes as well as healthy fats, such as nuts and olive oil, which have numerous health benefits while also supporting environmental sustainability. A study in Spain found that a Mediterranean diet resulted in a 72 percent decrease in greenhouse gas emissions, a 58 percent decrease in land use, a 52 percent reduction in energy usage, and a 33 percent decline in water consumption.[8]

Although the Mediterranean diet is often touted as one of the healthiest and most sustainable diets in the world, relatively few people actually practice this style of eating. When I lived in Italy, I noticed that very few people practiced the Mediterranean diet, despite that being one of its places of origin. I tend to call the Mediterranean diet the "disappearing diet," in that very few adhere to it these days. Soda replacing wine at the table in Rome and French fries and hotdogs as popular toppings for pizza in Naples are two examples in modern Italian diets that don't fit within the Mediterranean plan. There are other traditional, territorial diets that historically showed potential health benefits such as the Okinawan. These southwestern Japanese islanders are famous for living to be centenarians. Their diets of high amounts of vegetables (including sweet potatoes, shiitake mushrooms, and bitter melon), along with some fish, soy, and very little sugar and dairy, are credited in part for their remarkable life spans. Unfortunately, this diet has essentially disappeared. On Okinawa, noncommunicable diseases are skyrocketing due to changes in the diet, including the density of fast-food restaurants on the island, an impact, in part, of the American military base there. In general, this transition reflects the move away from "traditional diets" as globalization and urbanization spread across the planet.

In their 2016 review of 210 scenarios extracted from 63 studies, Lukasz Aleksandrowicz and his colleagues at the London School of Hygiene and Tropical Medicine found that vegan diets were associated with the greatest reductions in greenhouse gas emissions and in land use, while vegetarian diets were associated with the greatest reductions in water use.[9] Research on the carbon footprints associated with different dietary patterns have determined that shifting toward a vegan diet would reduce greenhouse gas emissions by 24 to 53 percent, shifting to a vegetarian diet would reduce emissions by 18 to 35 percent, and shifting to a Mediterranean diet would reduce emissions by 6 to 17 percent.[10]

PROMOTING FOOD PRACTICES
AND CONSCIOUSNESS

Food choice is not the only thing that determines health and environmental sustainability. A food practice can be divided into three distinct but interconnected practices: (1) the purchasing of food, (2) the preparing and cooking of food, and (3) the aesthetic judgments and taste in the consumption of food. Each of these practices can contribute to health and environmental outcomes.

When we as consumers bring our aspirations, values, and expectations to food environments, we can shift demand and influence the way food producers and suppliers behave. Introducing new conventions into a food system can also inspire the creation of new food "assemblages" that are more sustainable and healthier. In other words, food practices can enrich the diversity or supply of foods, thus giving people access to a greater variety of food environments.

Eating is not a list of dos and don'ts. It should be a pleasurable and rewarding experience. Mealtimes are important opportunities for socializing and building relationships. Traditional and cultural preferences in food choices need to be respected.

Food consciousness is the collection of activities that individuals can exercise to make better choices related to all aspects of their food environments. These activities include reducing food waste, paying more attention to daily spending and consumption, reading food labels, choosing sustainable packaging and "ugly," or imperfect foods, and being aware of the environmental implications of consuming certain foods, including how far they travelled to reach you.

One good way to practice food consciousness is by paying greater attention to food waste. Food waste tends to be invisible and thus is much less affected by social norms or social signaling. People may not realize how much food they throw away and how they could take action against food waste. One study in the Philippines found that the main reasons for wasting fruits and vegetables at the household level are forgetting to cook the produce purchased, not planning meals properly, and overbuying.[11] The disposal of food and drink waste in landfills adds to the release of greenhouse gases such as methane and wastes nutrients and resources that many people badly need.

New technologies to reduce food waste could help us practice food consciousness. Examples include in-store and online shopping lists, intelligent indicators of freshness or ripeness, smart refrigerators that allow the remote observation of food, and apps that enable the tracking of food freshness and waste. Although the use of such technologies would likely be skewed toward wealthier consumers, these are generally the consumers who generate the most waste.

Food production based on local knowledge, culture, and values can lead to the revival of nutritious traditional diets, offering consumers healthier choices. Traditional foods, apart from being vehicles of our cultures, may also possess health qualities, since tradition rarely honors foods that are not palatable and healthy. In Mali, a cooperative of female agroecological farmers, COFERSA (Convergence des Femmes Rurales pour la Souverainte Alimentaire), is creating new markets for their products by raising awareness about the nutritional benefits as well as the wonderful taste and texture of local foods, such as fonio, millet, and sorghum. I've had the pleasure of working with Pierre Thiam, a Senegalese chef and New York–based restaurateur, who started Yolele Foods. He promotes the importance of West African cuisine and particularly the supergrain fonio in how it can be prepared and consumed.

Globally important agriculture heritage systems also embody practices that can improve food systems. These systems don't differentiate between the environment and the people; they perceive that living beings and territory are interconnected and embedded with spirituality. This holistic view doesn't place humankind or the production of food at the center of the food system. Instead, maintaining the equilibrium between the environment and the beings inhabiting it is the central focus.

Many traditional populations have extensive understanding of nutritional issues from the field to the plate, including knowledge of the health and nutrition qualities of indigenous crop varieties. Knowledge of food preparation, combinations, processing, and preservation are an important part of the biocultural knowledge of many communities. Fermented foods are a good example. An in-depth review of indigenous peoples' food systems found that pride in local culture was one of the intervention strategies that improved health and nutrition among native peoples.[12] All of this informs their individual choices. However, the migration of youth to urban centers in search of education and job opportunities is threatening the intergenerational knowledge transmission fundamental for the survival of these traditional food systems. Supporting and preserving this knowledge among rural and indigenous peoples is critical. If we don't, we'll lose it forever.

Ultimately, changing our behaviors will be essential if we want to improve the nutrition of populations and preserve the environment. Behavioral change depends on many factors beyond self discipline, including education, knowledge, race, social

How Can We Reduce Food Waste?

Estimates indicate that up to a third of food is wasted across the planet, contributing to landfills and pollution from incineration and greenhouse gas emissions. Food waste includes what's thrown away or spoiled during production or preparation and what's left after consumption, including packaging. Food producers, stores, and restaurant owners have their own tools to lower waste percentages but here are things consumers can do to help:

- Buy only what's needed so food doesn't spoil before its "use by" date.
- Reduce portion sizes.
- Enroll in local Community Supported Agriculture programs (CSAs) to bring in fresh produce from nearby farms.
- Find ways to use all of what's edible, such as the stems of broccoli, and brown, spotted bananas, which are perfect for baking bread and muffins.
- Be creative in your cooking; modify recipes to use what you have on hand and reuse typically discarded by-products (like the bones from roasted chickens) for broths, sauces, etc.
- Compost eggshells, coffee grounds, and rinds to fertilize gardens and yards.
- Share surplus food with neighbors, friends, and local shelters as permitted.
- Purchase less-than-perfect produce from companies such as Misfit Market and Imperfect Foods.
- Recycle packaging—but be aware of what your local collection company will and won't pick up, and understand that much recycling ends up in landfills and waterways in Asia and most recently, Kenya, significantly decimating these landscapes and adding to the products' total carbon footprints through fossil fuel–driven transport.
- Download apps that guide you toward more sustainable grocery shopping and eating and connect you with restaurants that sell surplus portions at a discount, such as Food for All in Boston and New York and Too Good to Go in Europe and in some parts of the United States.

standing, mental health, stress levels, autonomy, control of resources, and social support from family and the community. New approaches to behavioral change that are sensitive to equity, social norms, and the cultural environment, grounded with some realism, will be essential to transform dietary habits.

IMPROVING FOOD LITERACY AND CULINARY EXPERIENCES

Transforming the global food system will require a universal understanding that what we eat matters, not only for individual well-being but also for the local community, the global community, and the planet. Individuals need better information about what constitutes a healthy diet and what actions they can take to support an equitable and sustainable food system.

Consumers are increasingly being asked to make complex choices about the food that they eat. The growing scientific complexity of food production and processing has placed greater burdens on consumers to understand food science. Media inundates consumers with messages about the health of our diets and the food system, but deciphering the science amid the "latest findings" is a challenge. Some messaging on nutrition, such as eating more fresh produce and consuming less salt, have largely been consistent across decades. But other advice about healthy, sustainable diets is more complicated or constantly shifting.

Nutrition education is important among individuals in all sectors of society—from people who are directly involved in health care and food production to consumers who make daily choices about what foods to eat. In particular, few health practitioners, or even medical students, receive nutrition education or training. Similarly, nutrition often falls through the cracks for community health workers and agriculture extension agents. Community health workers could screen and treat cases of malnutrition and provide nutrition education on health-related conditions on a regular basis. Health and extension agents could undergo nutrition training to provide knowledge to households on basic dietary guidelines and nutrition counseling. Joint trainings could enable collaboration across disciplines.

Beyond primary education, continued education, and technical training, academia can enhance public understanding of health and nutrition by bringing together farmers, producers, civil society representatives, youth, and others for dialogue, knowledge exchange, and capacity building. Smartphones and mobile technology can provide nutritional education and messages, allow for innovative participation in food markets, and reach remote nutritionally vulnerable households. Many opportunities exist within social media to increase understanding, transparency, and accountability. Parents, grandparents, employers, and employees can all be targeted with particular messages relevant to their responsibility and empower them to bring about changes in dietary habits. However, the question remains as how to best implement these technologies for diets and nutrition responsibly and how to ensure that social media platforms are not abused as marketing tools to sell unhealthy foods to young consumers.

Children and adolescents are potential trendsetters and taste makers. Education about healthy eating should be given more priority in schools, starting in preschool. As unhealthy, highly processed foods have become cheaper and more accessible and unprocessed foods such as vegetables and fish have become more expensive, ultra-processed foods are becoming a larger part of children's diets. In addition, the pleasurable colors, tastes, and textures of these foods appeal to children, fueled by aggressive marketing tactics. Factors that make these foods appealing should be considered in designing strategies to promote healthy foods for children. Introducing wholesome meals at school, coupled with educational programs, could establish children's eating practices early on, and they could then bring these new practices home to their families.

Food literacy is based on building an appreciation and understanding of the social, cultural, and environmental dimensions of food alongside practical food skills. Unfortunately, food literacy remains low in many regions of the world. When we were investigating why Americans valued meat so much (see pages 103–104), we found widespread confusion among consumers about which foods are healthy and sustainable. As part of the project, we asked consumers to sort different foods in any way they think they should be grouped. Participants ended up dividing foods into groups such as fruits, vegetables, and dairy. However, most consumers didn't know how to sort plant-based "milks" and meat alternatives, such as the Impossible Burger or Beyond Meat. Interestingly, they didn't consider them as part of any typical food group. In addition, most people still don't understand (or trust) lab-grown or cultured meats. The reaction to lab-grown meats will likely be similar to the often negative reactions to and suspicions of genetically modified foods when they first came on the market. These gaps in literacy of emerging technologies hinder the successful implementation of products that could have environmental and health benefits.

Raising awareness and educating people about food planning, purchasing, and handling can have positive effects on consumer attitudes and behaviors. Most of us are not aware of the freshness of the produce that we purchase in the grocery store. Potatoes are generally stored the longest, often held for up to four months before being sold. Once produce arrives at the store, it may sit on the shelves for days or weeks, especially if turnover is low. Though we can't control the age of the produce offered at the grocery store, we can buy fresher produce directly from farmers at growers-only markets, through Community Supported Agriculture shares, or through imperfect produce companies that may deliver produce and other items to remote areas. Unfortunately, farmer- and community-supported markets are not available everywhere, and if they are, you may get items that are unfamiliar, and the produce tends to be cheaper at the supermarket chains.

Produce freshness is important because fruits and vegetables have the highest nutritional value right after they're picked, with nutrient losses occurring largely through heat, light, and oxygen exposure. Even when produce is stored in environments optimized for preservation, losses still occur. One study examined 19 fruits and vegetables and found that, after 15 days in refrigeration many had decreased levels of vitamin C and antioxidant activity.[13] Frozen foods are a great second alternative, typically better than canned.

The date labeling on packaged foods is similarly important to understand. A "best before" date indicates the date when the food retains its expected quality; food can still be consumed past this date. A "use by" date refers to the date after which the food is not safe to be consumed. This information, combined with information on safe food handling at the household level, can both improve health and reduce consumer-level food waste.

Food preparation and cooking is another avenue through which to practice mindful food consciousness. Using safe and efficient methods of food preparation, such as energy-efficient cookstoves, can reduce the environmental impact of food preparation. Preparation techniques can also benefit health. Ensuring that foods rich in fat-soluble vitamins are cooked with oils can enhance the absorption of those vitamins. Cooking with iron cookware can improve iron absorption, particularly if acid-ic foods are cooked in iron pots at high temperatures. Acidic foods are often high in vitamin C, which can also enhance iron absorption. Cooking leafy green vegetables in minimal water and reusing the water (to include in soups, for instance) can ensure that water-soluble vitamins contained in those vegetables are not lost through the cooking process. Minimal milling of grains ensures that the nutritious parts of the seed are retained during consumption. Dehulling, peeling, soaking, germinating, fermenting, and drying certain foods can remove compounds that bind up micronutrients (such as phytates) and at the same time, preserve key micronutrients. These techniques are great, but they may be time consuming, and require investment and skills up front. Many people would love to experiment in the kitchen but they just don't have time or perhaps an equipped kitchen or, in some cases, running water or electricity.

Culinary knowledge and food skills are essential to healthy diets and good nutrition, but with the increased reliance on processed and prepared foods, people have lost their culinary chops. Sometimes it's just easier to walk into a market, grab, and go. Shopping and cooking take time and effort, and many people in the world struggle with time poverty. We shouldn't penalize people for not spending time in their kitchen. Improving culinary skills is one avenue, but that doesn't mean that food environments are off the hook. Perhaps for those of us who had to stay at home and socially distance during the

COVID-19 pandemic, the additional cooking we added to our days may become habit.

MOBILIZING FOR CHANGE

We can vote not only with our forks but with our dollars. We can support politicians and policies that promote better nutrition and sustainability. As human rights advocate Isatou Jallow has said, "Political will plus people's will equals sustainable will."[14]

Policy making is not only the domain of government. Civil societies can forge policy from their respective quarters as well. In France, the work of a local-level, grassroots initiative called AMAP (*Association pour le maintien d'une agriculture paysanne* [Association for Maintaining Small-Scale Family Farming]) successfully encouraged a shift in food practices. This association was born from the idea of enabling smallholders to keep their businesses alive through strong support from consumer groups and a focus on risk sharing. Such grassroots activism could generate even further support by linking the various movements related to nutrition and by forming coalitions and networks that work together to produce change.

Movements within our communities can also inspire change at the regional, national, or even global level. Social movements and civil society organizations can give voice to smallholder farmers, pastoralists, agricultural and food workers, small fisheries, foragers, native peoples, landless people, rural women, and young people around the world.

With the rise of digital technology, organizations, institutions, and individuals are now able to express their views immediately to a global audience. These platforms can be used for debates, education, capacity building, accountability, and monitoring.

The "1,000 Days" campaign, which emerged from *The Lancet*'s 2008 and 2013 series on maternal and child undernutrition, spread awareness that the first 1,000 days in a child's life is the critical window of opportunity to make a difference in their future. 1,000 Days is an advocacy hub that continues to champion new investment and partnerships by advocating for greater action, building investment in maternal and child nutrition, and catalyzing partnerships among different sectors to scale up efforts to reduce malnutrition. 1,000 Days is a slogan that people can wrap their minds around, and influential political figures such as former US Secretary of State Hillary Clinton brought widespread attention to the issue. The success of 1,000 Days prompted more than 60 countries to commit to scaling up nutrition programs.

Another way to mobilize communities is to get involved in local food systems. As individuals, we can get to know our local farmers, producers, and politicians. Developing relationships with the local people and organizations involved in food production systems can help us better understand and support their struggles. The promotion of healthy diets based on the local, seasonal production of foods, along with the promotion of short, nearby food distribution chains, can forge closer ties among farmers, consumers, and the land. The development of local distribution hubs — especially for healthy, fresh, and perishable products — could also reduce food waste from transport and consumption.

Communities can forge their own models and plans for health outside of government. An example of an innovative community-led public health model used to address acute malnutrition in developing countries is Community Management of Acute Malnutrition (CMAM), pioneered by Steve Collins of Valid Nutrition. This approach engages the community to detect early signs of severe acute malnutrition by sensitizing communities and encouraging active case finding. Originally, CMAM was used primarily in emergency settings. However, this approach was soon shown to be useful in nonemergency settings when the right components were in place. In Sauri, Kenya (near to the home of President Barack Obama's grandmother, who I had the pleasure of meeting while working in Sauri) community health workers and clinical staff were trained in this program as a way to prevent death among the 5 percent of young children who experienced acute malnutrition in the village. Through community awareness, the program diagnosed children suffering from acute malnutrition and referred them to one of three treatment modalities: (1) inpatient clinical therapeutic feeding,

(2) outpatient therapeutic feeding with specialized ready-to-use foods, and (3) household supplementary feeding with local foods. Other diseases that could be integrated into this platform include malaria, diarrhea, and other serious primary pediatric illnesses. The key piece of CMAM is community awareness to detect the early signs of malnutrition that then triggers the system of care and treatment for these children.

ADVOCATING FOR CHANGE

Individual leaders need to advocate for better nutrition and sustainability while also providing accurate information so that the public can make informed decisions. The nutrition community also needs diverse leaders, including scientists, advocates, writers, cooks, taste makers, and grassroots policy makers. The writer Michael Pollan, for example, is not a nutritionist or scientist, but has educated many people around the world about nutrition. His advocacy and books—including *The Omnivore's Dilemma: A Natural History of Four Meals* (2006), *In Defense of Food: An Eater's Manifesto* (2008), and *Cooked: A Natural History of Transformation* (2013)—have brought much greater awareness to issues of health and sustainability. "Eat food, mostly plants, not too much."[15] It doesn't get more digestible than that! While Pollan gets most of the food system challenges right, he often writes about nutrition issues in an idealized sense of what the food system should look like without necessarily addressing these issues for those living with more modest economic realities.

Another example of a global leader outside the science community is teenager Greta Thunberg, a Swedish environmental activist who has brought widespread attention to the issues involved in climate change. In 2019, Thunberg led coordinated multicity climate protests that involved millions of students. Her actions, which have ranged from individual changes to national efforts to global advocacy, have inspired millions more to follow in her footsteps. This combination of individual change, advocacy, and action is what's needed if we're to make our food systems healthy, equitable, and sustainable in the future.

A lot of innovative work is coming out of low-income countries. Young thought leaders in many low- and middle-income countries are shaping not only the food system but also how their countries are governed. Many young people want to disrupt the systems that have resulted in our current quandaries. Their intelligence and energy need to be harnessed to create opportunities for real change. During my time in East Timor, I met a young woman named Alva Lim who started a restaurant in the capital city of Dili where she employs about 15 Timorese youth who were previously unemployed. Her employees learn about the traditional foods of Timor-Leste, about how to prepare these foods as well as international cuisine, and about how to run a modern restaurant. Her efforts spread the word about local foods in Timor and encourage the Timorese to take pride in their food culture while also exposing them to new food trends.

Civil society organizations and communities of practice have increasingly proved their ability to mobilize, advocate, and launch initiatives to raise awareness and fight hunger both at the grassroots level and as a collective force at the regional and international levels. They're the spark that can urge governments to make changes. Within the food and nutrition sector, they have a particularly crucial role to play in supporting political processes and decision making, promoting sustainability-related issues at the institutional level, and raising awareness for sustainable diets among stakeholders, including youth.

Although the impact of grassroots advocacy may be limited when faced with such massive, global challenges, these disparate actions compound to produce the types of widespread change required to improve health and environmental outcomes. Governments and industries are intended to serve the needs of the people, and changes in individual beliefs should dictate policies, products, and food environments. Creating lasting, large-scale change will require a strengthened relationship between individual efforts and system-level changes. One cannot be effective without the other.

SO, *CAN* FIXING DINNER FIX THE PLANET?

Preventing catastrophic collapses within global food systems will require an all-hands-on-deck approach. On both individual

and systems levels, we need to be bolder. Our world is changing rapidly; we don't have time to let problems fester. No one
country can address climate change by itself. No one country
can steer food systems in the right direction. These are collective
world issues. As the saying goes, we're all in this together.

Governments need to make decisions now and be less risk
averse. Much evidence already exists about how to improve
food systems and diets, support climate adaptation, and drive climate mitigation. Strategies exist to address all these challenges
simultaneously. But for these approaches to be effective, governments need to commit to and invest in change, the private sector
has to participate and develop partnerships with other sectors
to improve public health and environmental sustainability, consumer awareness needs to increase, and young innovators need
to be supported along with more established inventors to bring
new ideas to the table. Citizens need to vote for leaders who will
foster global cooperation and goodwill.

Although an incredible amount of innovation is occurring
around the world, we need UN agencies, multilateral organizations, and governments to step into the twenty-first century,
be more nimble, and take responsibility for food systems that
have gone off the rails and put them back on track. International
development and the ways in which it functions are ripe for a
paradigm shift. New ways to think about these grand challenges
are emerging, and they're unlikely to come from just the UN or
World Bank. A different kind of architecture must push forward

an agenda to mitigate climate change and end poverty, hunger, and malnutrition in a way that none of us can currently predict.

World leaders established the Sustainable Development Goals at the UN General Assembly in 2015, creating a road map for sustainable development, but teens and people in their twenties are asking whether these plans are going to be enough. They have the opportunity and will to shift the agenda on climate and world hunger and rethink strategies moving forward. Already, events such as the Global Climate Strike in 2019 with the marchers chanting "the kids are not all right," have shown that this next generation will not stand by and do nothing. The Black Lives Matter protests are calling for an end to structural, institutional, and systemic racism and inequities, and city governments are listening—and actions are being taken. So much untapped talent exists in places like South Asia and sub-Saharan Africa and the youth living and working in these regions must be in the driver's seat to innovatively tackle global issues because they will be the most populated places on Earth. Our world will be reshaped by a younger generation that cares deeply about human and environmental health—and ultimately has more at stake from the consequences of climate change, as they'll live to see them. It's their food system we all must fight for.

Every country has problems that need to be fixed. Every country has some form of malnutrition. Every country will struggle with climate change, although some will struggle more than others. While the challenges may seem daunting, throughout history, people and nations have triumphed over great adversity, and miraculous human accomplishments have been born from struggle. Working together, I'm optimistic that food systems can adapt to support planetary and human health, and we can indeed bend the arc toward a better world that prioritizes equity and social justice. Seeing the writing on the wall, young people are leading the charge on many of these fronts, not just witnessing and talking about the problems, but also organizing and acting. More than anything, that gives me hope for our planet.

Acknowledgments

THIS BOOK WOULD NOT BE POSSIBLE without my incredible colleagues, postdocs, students, and teams, who've done a lot of the work highlighted in this book. Nor would it exist without the support of Michael R. Bloomberg, Johns Hopkins University's Office of Research, and my colleagues at the Schools of Public Health and Advanced International Studies and the Berman Institute of Bioethics.

My thanks go to Matthew R. McAdam and Robin Cruise who read the text and made useful comments on it and to my editor, Anna Marlis Burgard, who suffered through many food systems lectures and Zoom calls amid a pandemic to get the nitty-gritties right. And a massive thanks to Sarah Olson and Steve Olson who left their blood, sweat, and tears on earlier versions of the manuscript.

One is lucky if their mentors can be counted on two hands. I have many, but I must mention a few who have provided guideposts along the road, including Elaine Gallin, Sonia Sachs, Richard Deckelbaum, Glenn Denning, Cheryl Palm, Pedro Sanchez, C. J. Jones, Emile Frison, Pablo Eyzaguirre, Anna Lartey, Bjorn Ljunqvist, Martin Bloem, Ruth Faden, Jeff Kahn, and Lawrence Haddad. I also need to thank Roseline

Remans, Shauna Downs, So Young Jang, Gwyn Kirkbride, Leslie Engel, Mark McCormick, Swetha Manohar, Corinna Hawkes, Elizabeth Fox, Rebecca McLaren, Claire Davis, Quinn Marshall, Anna Herforth, Ahmed Raza, Anne Barnhill, Chris Bene, Fabrice DeClerck, and the Millennium Villages crews in New York and Africa who have worked with me over the last several years. Their partnership and friendship give every day meaning to the work we do.

I'm grateful to my parents for all their love and support, and to my sister, Kim, who has always been my biggest cheerleader. I'm so indebted to the many friends, colleagues, and fellow scientists I've named, and those I haven't named. Above all, I am grateful to Derek, my better half and partner in, well, everything.

This book would also not be possible without the people around the world who have participated in these studies and imparted their grassroots wisdom, including the farmers, their families, and on the ground health workers. Finally, we're indebted to those who have dealt with the consequences of poor nutrition and marginalization who've shared their stories to help us better understand their obstacles.

Notes

INTRODUCTION. YES, WE'LL HAVE NO BANANAS

1. Simon L. Lewis and Mark A. Maslin, "Defining the Anthropocene," *Nature* 519, no. 7542 (2015): 171–180.

2. Alex de Waal, *Mass Starvation: The History and Future of Famine* (n.p.: John Wiley & Sons, 2017).

3. Derek Byerlee and Jessica Fanzo, "The SDG of Zero Hunger 75 Years On: Turning Full Circle on Agriculture and Nutrition," *Global Food Security* 21 (June 2019): 52–59.

4. Food and Agriculture Organization et al., *The State of Food Security and Nutrition in the World 2020: Transforming Food Systems for Affordable Healthy Diets* (Rome: Food and Agriculture Organization, 2020).

5. Food and Agriculture Organization, *State of Food Insecurity*.

6. Ove Heogh-Guldberg et al., "2018 Impacts of 1.5°C Global Warming on Natural and Human Systems," in *Global Warming of 1.5 Degrees C: An IPCC Special Report on the Impacts of Global Warming of 1.5 Degrees C above Pre-industrial Levels and Related Global Greenhouse Gas Emission Pathways, in the Context of Strengthening the Global Response to the Threat of Climate Change, Sustainable Development, and Efforts to Eradicate Poverty*, ed. Valérie Masson-Delmotte et al. (Geneva, Switzerland: World Meteorological Organization, 2019).

CHAPTER 1. ARE WE WHAT WE EAT, OR WHAT WE'RE FED?

1. Jessica Fanzo and Rebecca McLaren, "Poor Countries Can't Live on Rice Alone," *Bloomberg Opinion*, 2017, https://www.bloomberg.com/opinion/articles/2017-06-12/poor-countries-can-t-live-on-rice-alone.

2. João Boavida, interview by the author.

3. GBD 2017 Diet Collaborators, "Health Effects of Dietary Risks in 195 Countries, 1990–2017: A Systematic Analysis for the Global Burden of Disease Study 2017," *The Lancet* 393, no. 10184 (2019): 1958–1972.

4. GBD 2017 Diet Collaborators, "Health Effects of Dietary Risks in 195 Countries."

5. Ashkan Afshin, interview by the author.

6. Enock Musinguzi, interview by the author.

7. Shaohua Chen and Martin Ravallion, "The Developing World Is Poorer Than We Thought, but No Less Successful in the Fight against Poverty," *Quarterly Journal of Economics* 125 (2010): 1577–1625.

8. Alisha Coleman-Jensen et al., *Household Food Security in the United States in 2018* (Washington, DC: Department of Agriculture, Economic Research Service, 2019).

9. Liping Pan et al., "Food Insecurity Is Associated with Obesity among Adults in 12 States," *Journal of the Academy of Nutrition and Dietetics* 112, no. 9 (2012): 1403–1409.

10. Lauren E. Au et al., "Household Food Insecurity Is Associated with Higher Adiposity among US Schoolchildren Ages 10–15 Years: The Healthy Communities Study," *Journal of Nutrition* 149, no. 9 (2019): 1642–1650.

11. Edward A. Frongillo and Jennifer Bernal, "Understanding the Coexistence of Food Insecurity and Obesity," *Current Pediatrics Reports* 2, no. 4 (2014): 284–290.

12. Max Roser and Hannah Ritchie, "Food Supply," *Our World in Data*, published online 2013, https://ourworldindata.org/food-supply.

13. A. Afshin et al., "Health Effects of Dietary Risks in 195 Countries, 1990–2017: A Systematic Analysis for the Global Burden of Disease Study 2017," *The Lancet* 393, no. 10184 (2019): 1958–1972.

14. Global Nutrition Report 2018.

15. Hannah Ritchie, "Meat Consumption Tends to Rise as We Get Richer," cited in "Meat and Dairy Production," *Our World in Data*, 2017, https://ourworldindata.org/meat-production#meat-consumption-tends-to-rise-as-we-get-richer.

16. S. E. Clark et al., "Exporting Obesity: US Farm and Trade Policy and the Transformation of the Mexican Consumer Food Environment," *International Journal of Occupational and Environmental Health* 18, no. 1 (2012): 53–64.

17. United Nations Children's Fund and World Health Organization, *Progress on Household Drinking Water, Sanitation, and Hygiene 2000–2017*, Special Focus on Inequalities (New York: UNICEF and WHO, 2019).

18. United Nations Children's Fund, World Health Organization, International Bank for Reconstruction and Development/The World Bank, *Levels and Trends in Child Malnutrition: Key Findings of the 2018 Edition of the Joint Child Malnutrition Estimates* (Geneva: WHO, 2019).

19. Lia C. Fernald and Lynnette M. Neufeld, "Overweight with Concurrent Stunting in Very Young Children from Rural Mexico: Prevalence and Associated Factors," *European Journal of Clinical Nutrition* 61, no. 5 (2006): 623–632.

20. Chris Hillbruner and Rebecca Egan, "Seasonality, Household Food Security, and Nutritional Status in Dinajpur, Bangladesh," *Food and Nutrition Bulletin* 29, no. 3 (2008): 221–231.

21. Robert E. Black et al., "Maternal and Child Undernutrition and Overweight in Low-Income and Middle Income Countries," *The Lancet* 382, no. 9890 (2013): P427–P451.

22. Lindsay Allen and Stuart R. Gillespie, *What Works? A Review of the Efficacy and Effectiveness of Nutrition Interventions* (Geneva in collaboration with the Asian Development Bank, Manila: ACC/SCN, 2001).

23. Swetha Manohar, interview by the author.

24. Marie Ng et al., "Global, Regional, and National Prevalence of Overweight and Obesity in Children and Adults during 1980–2013: A Systematic Analysis for the Global Burden of Disease Study 2013," *The Lancet* 384, no. 9945 (2014): 766–781.

25. NCD Risk Factor Collaboration, "Worldwide Trends in Diabetes since 1980: A Pooled Analysis of 751 Population Based Studies with 4.4 Million Participants," *The Lancet* 387, no. 10027 (2016): 1513–1530.

26. Maximillian Tremmel et al., "Economic Burden of Obesity: A Systematic Literature Review," *International Journal of Environmental Research and Public Health* 14, no. 4 (2017): 435.

27. Laura Michelle Cox and Martin J. Blaser, "Pathways to Microbe-Induced Obesity," *Cell Metabolism* 17, no. 6 (2013): 883–894.

28. Shauna M. Downs et al., "The Interface between Consumers and Their Food Environment in Myanmar: An Exploratory Mixed-Methods Study," *Public Health Nutrition* 22, no. 6 (2019): 1075–1088.

29. B. M. Popkin, C. Corvalan, and L. M. Grummer-Strawn, "Dynamics of the Double Burden of Malnutrition and the Changing Nutrition Reality," *The Lancet* 395, no. 10217 (2020): 65–74.

30. Micronutrient Initiative, *Investing in the Future—a United Call to Action on Vitamin and Mineral Deficiencies: Global Report 2009* (Ontario, Canada: United Call to Action, 2009).

31. World Health Organization, *The Global Prevalence of Anemia in 2011* (Geneva: WHO, 2015).

32. Development Initiatives, *2018 Global Nutrition Report: Shining a Light to Spur Action on Malnutrition* (Bristol, UK: Development Initiatives, 2018).

33. B. A. Swinburn et al., "The Global Syndemic of Obesity, Undernutrition, and Climate Change: The Lancet Commission Report," *The Lancet* 393, no. 10173 (2019): 791–846.

CHAPTER 2. CAN COOKING CURRY IN CAMBODIA TRIGGER A TORNADO IN TEXAS?

1. Intergovernmental Panel on Climate Change, *The Fifth Assessment Report of the IPCC: Synthesis Report* (Geneva, Switzerland: IPCC, 2014).

2. William J. Ripple et al., "World Scientists' Warning of a Climate Emergency," *BioScience* 70, no. 1 (2020): 8–12.

3. Colin K. Khoury et al., "Increasing Homogeneity in Global Food Supplies and the Implications for Food Security," *Proceedings of the National Academy of Sciences* 111, no. 11 (2014): 4001–4006.

4. Colin K. Khoury, interview by the author.

5. C. B. d'Amour et al., "Future Urban Land Expansion and Implications for Global Croplands," *Proceedings of the National Academy of Sciences* 114, no. 34 (2017): 8939–8944.

6. Deepak Ray et al., "Recent Patterns of Crop Yield Growth and Stagnation," *Nature Communications* 3, no. 1 (2012): 1–7.

7. Ove Heogh-Guldberg et al., "2018 Impacts of 1.5°C Global Warming on Natural and Human Systems," in *Global Warming of 1.5 degrees C: An IPCC Special Report on the Impacts of Global Warming of 1.5 Degrees C above Pre-industrial Levels and Related Global Greenhouse Gas Emission Pathways, in the Context of Strengthening the Global Response to the Threat of Climate Change, Sustainable Development, and Efforts to Eradicate Poverty*, ed. Valérie Masson-Delmotte et al. (Geneva, Switzerland: World Meteorological Organization, 2019).

8. F. Gaupp et al., "Changing Risks of Simultaneous Global Breadbasket Failure," *Nature Climate Change* 10, no. 1 (2020): 54–57.

9. Johan Rockström, interview by the author.

10. Gaupp et al., "Changing Risks," 54–57.

11. David Tilman and Michael Clark, "Global Diets Link Environmental Sustainability and Human Health," *Nature* 515 (2014): 518–522.

12. Jessica Fanzo et al., "The Effect of Climate Change across Food Systems: Implications for Nutrition Outcomes," *Global Food Security* 18 (2018): 12–19.

13. Karl Gruber, "Agrobiodiversity: The Living Library," *Nature* 544, no. 7651 (2017): S8–S10.

14. Luigi Guarino, "How Many Rice Varieties Are There in India?," *Agricultural Biodiversity Weblog*, March 11, 2020, https://agro.biodiver.se/2020/03/how-many-rice-varieties-are-there-in-india/.

15. Khoury et al., "Increasing Homogeneity in Global Food Supplies," 4001–4006.

16. Mario Herrero et al., "Farming and the Geography of Nutrient Production for Human Use: A Transdisciplinary Analysis," *Lancet Planetary Health* 1, no. 1 (2017): e33–e42.

17. Pedro A. Sánchez et al., *Halving Hunger: It Can Be Done*, Summary Version of the Report of the Task Force on Hunger (New York: Earth Institute at Columbia University, 2005).

18. Matthew Ryan Smith et al., "Effects of Decreases of Animal Pollinators on Human Nutrition and Global Health: A Modeling Analysis," *The Lancet* 386, no. 10007 (2015): 1964–1972.

19. United Nations Convention to Combat Desertification, "Threats to Soils: Global Trends and Perspectives" (Global Land Outlook Working Paper, a contribution from the Intergovernmental Technical Panel on Soils, Global Soil Partnership, Food and Agriculture Organization of the United States, 2017).

20. Food and Agriculture Organization of the United States, *Water and Food: The Post 2015 Water Thematic Consultation—Water Resources Management Stream Framing Paper* (Rome: FAO, 2013).

21. Food and Agriculture Organization of the United Nations, *Climate Change and Food Security: Risks and Responses* (Rome: FAO, 2016).

22. Christina C. Hicks et al., "Harnessing Global Fisheries to Tackle Micronutrient Deficiencies," *Nature* 574 (2019): 95–98.

23. William Cheung et al., "Large-Scale Redistribution of Maximum Fisheries Catch Potential in the Global Ocean under Climate Change," *Global Change Biology* 16, no. 1 (2010): 24–35.

24. National Academies of Sciences, Engineering, and Medicine, *Genetically Engineered Crops: Experiences and Prospects* (Washington, DC: National Academies Press, 2016).

25. Wilhelm Klümper and Matin Qaim, "A Meta-analysis of the Impacts of Genetically Modified Crops," *PLOS ONE* (November 3, 2014).

26. Graham Brookes and Peter Barfoot, "Farm Income and Production Impacts of Using GM Crop Technology 1996–2016," *GM Crops & Food: Biotechnology in Agriculture and the Food Chain* 9, no. 2 (2018).

27. E. D. Perry et al., "Genetically Engineered Crops and Pesticide Use in US Maize and Soybeans," *Science Advances* 2, no. 8 (2016): e1600850.

28. L. P. Agostini et al., "Effects of Glyphosate Exposure on Human Health: Insights from Epidemiological and In Vitro Studies," *Science of the Total Environment* 705 (2020): 135808.

29. Howdy Bouis, interview by the author.

30. Ruth S. DeFries et al., "Synergies and Trade-Offs for Sustainable Agriculture: Nutritional Yields and Climate-Resilience for Cereal Crops in Central India," *Global Food Security* 11 (2016): 44–53.

31. Glenn Denning and Jessica Fanzo, "Ten Forces Shaping the Global Food System," in *Good Nutrition: Perspectives for the 2st Century*, ed. K. Kraemer et al. (Basel, Switzerland: Karger, 2016), chap. 1.1, 19–30; M. F. Bellemare, J. Fajardo-Gonzalez, and S. R. Gitter, "Foods and Fads: The Welfare Impacts of Rising Quinoa Prices in Peru," *World Development*, 112 (2018): 163–179.

32. Jenny Gustavsson et al., *Global Food Losses and Food Waste—Extent, Causes, and Prevention* (Rome: FAO, 2011).

33. M. K. Albright, "The Moral Imperatives of Food Security," *Aspen Journal of Ideas*, May/June 2015, http://aspen.us/journal/editions/mayjune-2015/moral-imperatives-food-security.

CHAPTER 3. DO WE HAVE THE RIGHT TO EAT WRONGLY?

1. Tara Garnett, "Three Perspectives on Sustainable Food Security: Efficiency, Demand Restraint, Food System Transformation: What Role for Life Cycle Assessment?," *Journal of Cleaner Production* 73 (2014): 1–9.

2. Derek D. Headey and Harold H. Alderman, "The Relative Caloric Prices of Healthy and Unhealthy Foods Differ Systematically across Income Levels and Continents," *Journal of Nutrition* 149, no. 11 (2019): 2020–2033.

3. Food and Agriculture Organization et al., *The State of Food Security and Nutrition in the World 2020: Transforming Food Systems for Affordable Healthy Diets* (Rome: FAO, 2020).

4. HLPE, *Nutrition and Food Systems*, A Report by the High Level Panel of Experts on Food Security and Nutrition of the Committee on World Food Security (Rome: HLPE, 2017).

5. Joseph Stiglitz and Andrew Charlton, *Fair Trade for All: How Trade Can Promote Development* (Oxford: Oxford University Press, 2005).

6. Stephen Devereux, "Seasonality and Social Protection in Africa" (working paper, Future Agricultures, 2009).

7. Food and Agriculture Organization et al., *The State of Food Security and Nutrition in the World 2018: Building Climate Resilience for Food Security and Nutrition* (Rome: FAO, 2018).

8. Food and Agriculture Organization et al., *The State of Food Security and Nutrition in the World 2017: Building Resilience for Peace and Food Security* (Rome: FAO, 2017).

9.   Group of International and Regional Eminent Experts on Yemen, "Yemen: Collective Failure, Collective Responsibility," UN Expert Report, 2019, https://www.ohchr.org/EN/HRBodies/HRC/YemenGEE/Pages/Index.aspx.

10. Imogen Calderwood, "The Destruction of This One Port Could Cause Devastation for Yemen's Already Starving People," Global Citizen, May 31, 2018, https://www.globalcitizen.org/en/content/yemen-hodeidah-port-battle-famine-displacement/.

11. Allison Aubrey, "Dollar Stores and Food Deserts," *Sunday Morning*, December 8, 2019, https://www.cbsnews.com/news/dollar-stores-and-food-deserts-the-latest-struggle-between-main-street-and-corporate-america/.

12. Kristen Cooksey-Stowers, Marlene B. Schwartz, and Kelly D. Brownell, "Food Swamps Predict Obesity Rates Better Than Food Deserts in the United States," *International Journal of Environmental Research and Public Health* 14, no. 11 (2017).

13. Carlos A. Monteiro and Geoffrey Cannon, "The Impact of Transnational 'Big Food' Companies on the South: A View from Brazil," *PLOS Medicine* 9, no. 7 (2012): e1001252.

14. Marion Nestle, interview by the author.

15. J. Poore and T. Nemecek, "Reducing Food's Environmental Impacts through Producers and Consumers," *Science* 360, no. 6392 (2018): 987–992.

16. T. Adesogan et al., "Animal Source Foods: Sustainability Problem or Malnutrition and Sustainability Solution? Perspective Matters," *Global Food Security* 25 (2019): 100325.

17. Paul Wilkinson et al., "Public Health Benefits of Strategies to Reduce Greenhouse-Gas Emissions: Household Energy," *The Lancet* 374, no. 9705 (2009): 1917–1929.

18. J. Ranganathan et al., "Shifting Diets for a Sustainable Food Future" (working paper, Installment 11 of Creating a Sustainable Food Future, World Resources Institute, Washington, DC, 2016), http://www.worldresourcesreport.org.

19. Ranganathan et al., "Shifting Diets for a Sustainable Food Future."

20. Anne Mottet and Giuseppe Tempio, "Global Poultry Production: Current State and Future Outlook and Challenges," *World's Poultry Science Journal* 73, no. 2 (2017): 245–256.

21. Vernon H. Heywood, "Overview of Agricultural Biodiversity and Its Contribution to Nutrition and Health," in *Diversifying Food and Diets: Using Agricultural Biodiversity to Improve Nutrition and Health*, ed. Jessica Fanzo et al. (London: Earthscan/Routledge, 2013); Food and Agriculture Organization of the United Nations, *The State of the World's Animal Genetic Resources for Food and Agriculture*, ed. B. Rischkowsky and D. Pilling (Rome: FAO, 2007).

22. Jonathan A. Foley et al., "Solutions for a Cultivated Planet," *Nature* 478 (2011): 337–342.

23. FAO, *Climate Change and Food Security*.

24. Female herder, interview with the author.

25. Elizabeth L. Fox and Jessica Fanzo, "The Case of Pastoralism in Northern Kenya: Food, Water, Land and Livelihoods" (22nd International Congress of Nutrition, International Union of Nutritional Sciences, Buenos Aires, Argentina, 2017).

26. Rancher, interview with the author.

27. Worker, interview with the author.

28. Anne Mottet et al., "Climate Change Mitigation and Productivity Gains in Livestock Supply Chains: Insights from Regional Case Studies," *Regional Environmental Change* 17, no. 1 (2017): 129–141.

29. Food and Agriculture Organization of the United Nations, *The State of Food and Agriculture: Climate Change, Agriculture and Food Security* (Rome: FAO, 2016).

30. Food and Agriculture Organization of the United Nations, *The State of Food and Agriculture: Women in Agriculture; Closing the Gender Gap for Development* (Rome: FAO, 2011).

31. Collette Owens, Justine Dandy, and Peter Hancock, "Perceptions of Pregnancy Experiences When Using a Community-Based Antenatal Service: A Qualitative Study of Refugee and Migrant Women in Perth, Western Australia," *Women Birth* 29, no. 2 (2016): 128–137.

32. Lisa C. Smith et al., *The Importance of Women's Status for Child Nutrition in Developing Countries* (Washington, DC: International Food Policy Research Institute, 2003).

33. Corinna Hawkes and Marie T. Ruel, *From Agriculture to Nutrition: Pathways, Synergies, and Outcomes* (World Bank Other Operational Studies 9511, World Bank, Washington, DC, 2008).

CHAPTER 4. CAN BETTER POLICIES CREATE BETTER FOOD?

1. Dariush Mozaffarian et al., "Role of Government Policy in Nutrition—Barriers to and Opportunities for Healthier Eating," *BMJ* 361 (2018): k2426.

2. Parviz Koohafkan and Miguel A. Altieri, *Globally Important Agricultural Heritage Systems: A Legacy for the Future* (Rome: FAO, 2011).

3. Lois Englberger, *Let's Go Local: Guidelines Promoting Pacific Island Foods* (Rome: FAO, 2011).

4. FAO and GIAHS, "Globally Important Agriculture Heritage Systems," www.fao.org/giahs/en.

5. J. Fanzo et al., "Integration of Nutrition into Extension and Advisory Services: A Synthesis of Experiences, Lessons, and Recommendations," *Food and Nutrition Bulletin* 36, no. 2 (2015): 120–137.

6. Glenn Denning, interview by the author.

7. L. Klerkx and D. Rose, "Dealing with the Game-Changing Technologies of Agriculture 4.0: How Do We Manage Diversity and Responsibility in Food System Transition Pathways?," *Global Food Security* 24 (2020): 100347.

8. E. S. Cassidy et al., "Redefining Agricultural Yields: From Tonnes to People Nourished per Hectare," *Environmental Research Letters* 8, no. 3 (2013): 034015.

9. Rachel Bezner Kerr, Peter R. Berti, and Lizzie Shumba, "Effects of a Participatory Agriculture and Nutrition Education Project on Child Growth in Northern Malawi," *Public Health Nutrition* 14, no. 8 (2011): 1466–1472.

10. F. A. DeClerck et al., "Ecological Approaches to Human Nutrition," *Food and Nutrition Bulletin* 32, no. 1, suppl. 1 (2011): S41–S50.

11. FAO/IIRR/World Fish Center, *Integrated Agriculture-Aquaculture: A Primer* (Rome: FAO, 2001).

12. Andrew Martin and Kim Severson, "Sticker Shock in the Organic Aisles," *New York Times*, April 18, 2008.

13. "Comparison of SNAP Authorized Farmers and Markets FY2012 and FY2017," Food and Nutrition Service, https://fns-prod.azureedge.net/sites /default/files/snap/SNAP-Farmers-Markets-Redemptions.pdf.

14. Roland Sturm et al., "A Cash-Back Rebate Program for Healthy Food Purchases in South Africa: Results from Scanner Data," *American Journal of Preventive Medicine* 44, no. 6 (2013): 567–572.

15. R. H. Thaler and C. R. Sunstein, *Nudge: Improving Decisions about Health, Wealth, and Happiness* (New York: Penguin, 2009).

16. Walter Willett et al., "Food in the Anthropocene: The EAT-*Lancet* Commission on Healthy Diets from Sustainable Food Systems," *The Lancet* 393, no. 10170 (2019): 447–492.

17. W. Willett et al., "Our Food in the Anthropocene: The EAT-*Lancet* Commission on Healthy Diets from Sustainable Food Systems," *The Lancet* (2019): 1–47.

18. Kalle Hirvonen et al., "Affordability of the EAT-*Lancet* Reference Diet: A Global Analysis," *Lancet Global Health* 8, no. 1 (2020): eE59–eE66.

19. Mireya Valdebenito Verdugo et al., *Informe de resultados: descripcion de las percepciones y actitudes de los/as consumidores respect a las medidas estatales en el marco de la implementacion del Decreto 13/15*, Licitacion ID: 757-98-LQ16, 2017.

20. L. S. Taillie, "An Evaluation of Chile's Law of Food Labeling and Advertising on Sugar-Sweetened Beverage Purchases from 2015 to 2017: A Before-and-After Study," *PLOS Medicine*, 17(2): e1003015.

21. Ricardo Uauy, interview with the author.

22. C. T. Birt et al., "Healthy and Sustainable Diets for European Countries" (Report of a working Group, EUPHA, 2017).

23. M. Springmann et al., "The Healthiness and Sustainability of National and Global Food Based Dietary Guidelines: Modelling Study," *BMJ* 370 (2020).

24. "2015 Dietary Guidelines: Giving You the Tools You Need to Make Healthy Choices," USDA, https://www.usda.gov/media/blog/2015/10/06/2015 -dietary-guidelines-giving-you-tools-you-need-make-healthy-choices; *13th DGE-Nutrition Report: 2016 Summary* (Bonn, Germany: Germany Nutrition Society, 2016).

25. S. Ahmed, S. Downs, and J. Fanzo, "Advancing an Integrative Framework to Evaluate Sustainability in National Dietary Guidelines," *Frontier in Sustainable Food Systems* 3, no. 76 (2019): 10-3389.

26. Ministry of Health of Brazil, *Dietary Guidelines for the Brazilian Population* (Brasilia: Ministry of Health of Brazil, 2018), http://bvsms.saude.gov.br/bvs /publicacoes/dietary_guidelines_brazilian_population.pdf.

27. A. M. Thow et al., "Fiscal Policy to Improve Diets and Prevent Noncommunicable Diseases: From Recommendations to Action," *Bulletin of the World Health Organization* 96, no. 3 (2018): 201; A. Afshin et al., "The Prospective Impact of Food Pricing on Improving Dietary Consumption: A Systematic Review and Meta-analysis," *PLOS ONE* 12, no. 3 (2017).

28. M. Arantxa Colchero et al., "In Mexico, Evidence of Sustained Consumer

Response Two Years after Implementing a Sugar-Sweetened Beverage Tax," *Health Affairs*, 36, no. 3 (2017): 564–571.

29.   Stephen A. Wood et al., "Trade and the Equitability of Global Food Nutrient Distribution," *Nature Sustainability* 1, no. 1 (2018): 34–37.

30.   Lawrence Haddad, interview with the author.

31.   International Food Policy Research Institute, *Global Nutrition Report 2016: From Promise to Impact: Ending Malnutrition by 2030* (Washington, DC: IFPRI, 2016).

32.   Rachel Nugent and Andrea B. Fiegl, *Where Have All the Donors Gone? Scarce Donor Funding for Non-communicable Diseases* (Washington, DC: Center for Global Development, 2010).

33.   Meera Shekar et al., *An Investment Framework for Nutrition: Reaching the Global Targets for Stunting, Anemia, Breastfeeding and Wasting* (Washington, DC: World Bank, 2016).

34.   Shekar et al., *An Investment Framework for Nutrition*.

35.   Varsha Vijay et al., "The Impacts of Oil Palm on Recent Deforestation and Biodiversity Loss," *PLOS ONE* 11, no. 7 (2016): e0159668.

36.   Shauna Downs, interview by the author.

37.   Shauna M. Downs and Jessica Fanzo, "Is a Cardio-protective Diet Sustainable? A Review of the Synergies and Tensions between Foods That Promote the Health of the Heart and the Planet," *Current Nutrition Reports* 4, no. 4 (2015): 313–322; Davy Vanham, Mesfin M. Mekonnen, and Arjen Y. Hoekstra, "Treenuts and Groundnuts in the EAT-*Lancet* Reference Diet: Concerns regarding Sustainable Water Use," *Global Food Security* 24 (2020): 100357.

CHAPTER 5. CAN ONE BEE SAVE THE HIVE?

1.   Food and Agriculture Organization and World Health Organization, *Sustainable Healthy Diets—Guiding Principles* (Rome: FAO and WHO, 2019).

2. Stephen E. Clune, Enda Crossin, and Karli Verghese, "Systematic Review of Greenhouse Gas Emissions for Different Fresh Food Categories," *Journal of Cleaner Production* 140, no. 2 (2017): 766–783.

3. Kari Hamershlag, *Meat Eater's Guide to Climate Change and Health* (Washington, DC: Environmental Working Group, 2011); J. Poore and T. Nemecek, "Reducing Food's Environmental Impacts through Producers and Consumers," *Science* 360, no. 6392 (2018): 987–992.

4. Tim Searchinger, Richard Waite, Craig Hanson, and Janet Ranganathan, *Creating a Sustainable Food Future: A Menu of Solutions to Feed Nearly 10 Billion People by 2050* (Washington, DC: World Resources Institute, 2018).

5. Janet Ranganathan et al., *Shifting Diets for a Sustainable Food Future: Creating a Sustainable Food Future* (Washington, DC: World Resources Institute, 2016).

6. Marco Springmann et al., "Analysis and Valuation of the Health and Climate Change Cobenefits of Dietary Change," *Proceedings of the National Academy of Sciences* 113, no. 15 (2019): 4146–4151.

7. Lukasz Aleksandrowicz et al., "The Impacts of Dietary Change on Greenhouse Gas Emissions, Land Use, Water Use, and Health: A Systematic Review," *PLOS ONE* 1, no. 11 (2016): e0165797.

8. Sara Sáez-Almendros et al., "Environmental Footprints of Mediterranean versus Western Dietary Patterns: Beyond the Health Benefits of the Mediterranean Diet," *Environmental Health* 12 (2013): 118.

9. Aleksandrowicz et al., "The Impacts of Dietary Change," e0165797.

10. E. Hallström, A. Carlsson-Kanyama, and P. Börjesson, "Environmental Impact of Dietary Change: A Systematic Review," *Journal of Cleaner Production* 91 (2015): 1–11.

11. Elda B. Esguerra, Dormita R. del Carmen, and Rosa S. Rolle, "Purchasing Patterns and Consumer Level Waste of Fruits and Vegetables in Urban and Peri-urban Centers in the Philippines," *Food and Nutrition Sciences* 8, no. 10 (2017): 961–977.

12. Harriet V. Kuhnlein et al., *Indigenous Peoples' Food Systems and Well-Being: Interventions and Policies for Healthy Communities* (Rome: FAO, 2013).

13. Joseph H. Y. Galani et al., "Storage of Fruits and Vegetables in Refrigerator Increases Their Phenolic Acids but Decreases the Total Phenolics, Anthocyanins and Vitamin C with Subsequent Loss of Their Antioxidant Capacity," *Antioxidants (Basel)* 6, no. 3 (2017): 59.

14. M. Harper et al., "The Challenges of Sustainable Food Systems Where Food Security Meets Sustainability—What Are Countries Doing?," in *Sustainable Diets: Linking Nutrition and Food Systems*, ed. B. Burlingame and S. Dernini (Oxfordshire, UK: CAB International, 2019), chap. 3.

15. M. Pollan, *In Defense of Food: An Eater's Manifesto* (New York: Penguin, 2008).

Index